D1411278

BUGBUSTERS

BUGBUSTERS

GETTING RID OF
HOUSEHOLD
PESTS
WITHOUT
DANGEROUS CHEMICALS

MADISON COUNTY
CANTON PUBLIC LIBRARY SYSTEM
CANTON, MISS. 39046

648.7
Lif
c.1

Bernice Lifton

Foreword by Vincent L. Tofany
Former president of the National Safety Council

McGRAW-HILL BOOK COMPANY

New York   St. Louis   San Francisco   Bogotá   Guatemala   Hamburg
Lisbon   Madrid   Mexico   Montreal   Panama   Paris   San Juan
São Paulo   Tokyo   Toronto

The techniques discussed in this book are derived from reputable and reliable sources, but their efficacy is not guaranteed. Neither the author nor the publisher shall be responsible for their failure or misuse. The reader is advised to follow these techniques without deviation.

Copyright © 1985 by Bernice Lifton

All rights reserved. Printed in the United States of America. Except as permitted under the Copyright Act of 1976, no part of this publication may be reproduced or distributed in any form or by any means, or stored in a data base or retrieval system, without the prior written permission of the publisher.

1 2 3 4 5 6 7 8 9 DOCDOC 8 7 6 5

ISBN 0-07-037848-7 {PBK.}
ISBN 0-07-037849-5 {H.C.}

LIBRARY OF CONGRESS CATALOGING IN PUBLICATION DATA

Lifton, Bernice.
  Bugbusters: getting rid of household pests without dangerous chemicals.

  1. Household pests—Control.   I. Title.
TX325.L54 1985   648'.7   84-26182

ISBN 0-07-037849-5

ISBN 0-07-037848-7 (pbk.)

BOOK DESIGN BY PATRICE FODERO

To David

Contents

Acknowledgments

Many people have given generously of their expertise and time to the writing of this book. From distinguished scientists, busy health professionals, conscientious pest-control operators, and dedicated employees at all levels of government, I've enjoyed unstinting cooperation. Understanding the need for a book such as this, without exception they offered enthusiastic support. I'm grateful to all of them.

Professor Emeritus Walter Ebeling of the University of California at Los Angeles read the manuscript for technical accuracy, offering many pertinent suggestions. His comprehensive book *Urban Entomology,* a basic pest-control reference work, has been invaluable to me in this undertaking. James M. Stewart of the Centers for Disease Control in Atlanta was ever ready with source material, illustrations, imaginative research leads, and answers to my questions. Charles Meyers and Minoo Madon of the California Department of Health Services also answered many questions both in personal interviews and phone calls.

Others who found time to talk with me at length include Dr. Paul Cheo, formerly research director of the Los Angeles State and County Arboretum; Boyd Foster, president of Arrowhead Mills; Robert E. La-Voie, president of Ace Pest and Termite Control; Dr. George Peavey of the Southern California Veterinary Association; Drs. Susan Tully and Bernard Portnoy, pediatricians at the Los Angeles County–University of

Southern California Medical Center; Dr. George Rambo, technical director of the National Pest Control Association; Corinne Ray, poison information administrator of the Los Angeles County Medical Association; and Richard Yaussie, pest-control operator of Long Beach, California.

Most helpful also were Professor Robert Wagner of the University of California at Riverside; Dr. C. E. Schreck of the United States Department of Agriculture's Insect Repellent and Attractant Project; Pat Morrow, veterinary assistant of San Jose, California; James Boland, Jerome Blondell, Albert Heier, Pat Maravilla, and Carol Parker of the Environmental Protection Agency; Howard Dumont of the California Structural Pest Control Board; Lorraine Lawrence of the Consumer Product Safety Commission; Thomas Mulhern of the American Mosquito Control Association; and Thomas Reardon, environmental health coordinator for the city of Pasadena, California.

Warm thanks to my agents, Bart Andrews and Sherry Robb, who saw the idea's possibilities from the first and whose enthusiasm for the project never flagged.

A special bow to PJ Haduch, my editor at McGraw-Hill. With keen analytical sense, creativity, tact, and unending good humor, she helped me shape an unwieldy first draft into a finished book. A perceptive, gifted editor, she's just the friend an author needs.

Thanks, too, to Marg Hainer for a careful job of copyediting.

Lastly, loving thanks to my husband, who made sure I took needed breaks from my desk, reassured me when I worried, cheerfully adopted unfamiliar pest controls in our home and garden, and never doubted that I would finish this book.

Bernice Lifton
Pasadena, 1985

Foreword

No one likes finding a cockroach in the kitchen or seeing a pet in misery from fleas. These and other pests threaten our health and peace of mind, contaminate our food, make us sick, damage our clothes and homes, and ruin our gardens. It's no wonder we're anxious to clear them out in a hurry.

Modern chemistry seems to have given us a fast, convenient solution to our pest problems: Aim an aerosol can at the offender and watch it quiver, curl up, and die on the spot.

Or does it? Some pests (hundreds, in fact) are unaffected by the chemists' compounds. Others are so numerous or reproduce so quickly that replacements soon show up and we have to spray, bomb, or powder over and over again.

Children of technology that we are, we expect easy, immediate fulfillment of our needs—fast foods, fast information, fast transportation, and fast pest killers. Most of these are excellent solutions to our problems: A quick, reasonably nutritious meal when we're in a hurry, accurate data as soon as we need it, a 3000-mile trip in half a working day.

Chemical insecticides, however, can pose a danger when consumers ignore label warnings and handle these materials casually. Improperly used or carelessly stored, they can do us as much harm as do the pests. Thousands of people are treated every year in hospitals or emergency rooms for pesticide-related incidents. Thousands more contact poison information centers, complaining of nausea, dizziness, or headache after treating their homes for unwanted insects or rodents. Most of the victims are not farm laborers or chemical-factory workers. They are people—or their children—from all walks of life who are trying to hold down the number of household pests. They may be allergic to the chemicals or think that if a small amount of a chemical does a good job, twice the amount will do a better job. For such people the cure proves worse than the disease.

We seem to be in a blind alley with no exit. We don't want our environment polluted with pests and we don't want it polluted with chemicals. Actually, there *is* a way out. This is one instance where you can have your cake and eat it too. By following the strategies described in this book, you can have a home virtually free of unwanted wildlife while at the same time enjoying a safer personal environment.

Topping the list of safe antipest strategies is consistently careful home maintenance. Finding little food, no entryways, and no dusty corners in which to hide itself or its eggs, an insect won't hang around long. With equipment no harder to use than a vacuum sweeper, an oven, a refrigerator-freezer, a caulking gun, and a paint brush, you're well armed to oust most household pests, whether four- or six-legged. Simple traps and barriers along with sunshine and herbals known to repel insects complete your armory.

Home is meant to be our sanctuary, a haven of rest and comfort. It can also be a perilous place where we slip in the bathtub, fall off a stepladder, shock ourselves with frayed electric cords, or accidentally ingest toxic substances meant to keep our surroundings clean.

Among the easiest home accidents to prevent are those arising from the misuse of pesticides. The controls detailed in these pages vary from one animal to another, but nearly all are simple and easy to apply. No matter what small creature you're confronting, these should be the first means you use. Given a chance, they will cut your pest problem down to tolerable size. And your home will be a safer place.

Vincent L. Tofany
Former president
National Safety Council

Introduction

In 1981, across the United States 21,246 people—over half of them children—were treated for pesticide poisoning in hospital emergency rooms. More than 2000 of these victims were subsequently hospitalized, mostly either the very young or the very old. The vast majority had been poisoned at home, only 15 percent on farms or at workplaces. And the home-related accidents tended to be more serious. Four times more household victims were hospitalized than workplace victims.

Inevitably some victims have died. Between 1974 and 1976, nationwide, 120 people died in hospitals from pesticides; 87 died outside of hospitals in 1976.

Says Dr. Sue Tully, director of the pediatric emergency room at the Los Angeles County–University of Southern California Medical Center, "This is just the tip of the iceberg. You can figure that only about 10 to 15 percent of poisonings are reported."

Researchers at the Environmental Protection Agency (EPA) agree with Dr. Tully. They estimate that in 1976–1977 about 2.5 million U.S. households experienced nausea, dizziness, headache, or vomiting after using pesticides.

Considering how lavishly we consume these substances, the figures are not surprising. In 1979 better than 90 percent of all U.S. homes used pesticides and applied them indoors two to four times more frequently than outdoors.

We householders consume one-third of all pesticides sold in the United States, buying products with over 500 different formulations. We expose ourselves over and over to moth crystals, pesticide strips, roach and ant sprays, and pet insecticides. Many of the twenty most widely used antipest substances are suspected cancer agents. Most of us encounter these toxic materials on a chronic, long-term basis.

Despite the recent uproar set off by agricultural spraying of malathion, the home, concludes the EPA, may be the public's chief source of pesticide exposure.

This book is not a polemic against chemical pesticides. They are very necessary if Americans—and the rest of the world's people—are to have enough to eat. In some less developed societies, insects and rodents destroy more than 40 percent of the food crops. When modern farming, including the proper use of chemicals, is introduced, people once chronically on the verge of starvation, suddenly have enough food. Their well-being improves dramatically.

We can hope that as awareness of the risks in overuse of chemicals spreads, farmers will use these substances very sparingly. A growing number of them are following the techniques of Integrated Pest Management and using only the minimum amount of pesticides needed to bring in their crops. The rest of us would do well to follow their example.

Pesticides are something like drugs. There are problems, like termites, for which they are, at present, the best solution we have. But, like drugs, if we use too many too often, we need more and more of them to get results. They gradually become less effective and when we really need them, they may not work at all.

"Well," you ask, "if the government didn't know this stuff was safe, they wouldn't permit it to be sold, would they?"

Maybe. In May 1984 high officials of the nation's leading biochemical testing company, Industrial Bio-Test Laboratories, were found guilty of falsifying test results on four widely used biochemicals, among them an insecticide and an herbicide. Worthless tests like these had won the Food and Drug Administration's approval for 212 substances. Fifteen percent of all pesticides being sold in 1983 had been approved on the basis of at least some invalid tests done at the discredited lab. The consumer's blind faith in the idea it must be safe if it is sold hardly seems realistic.

Even if the tests are accurate and the government's approval reliable, there are more subtle dangers—dangers that may not surface for a long time. Very little is known about the interaction of small doses of pesticides in combination with drugs, alcohol, and food additives. We could find out—too late—many years from now.

When DDT was discovered, it was hailed as the ultimate answer to the world's pest problems. Typhus epidemics raging in Asia and postwar Europe were stopped cold; the malaria-carrying anopheles mosquito seemed about to join the dinosaurs. In 1948, Dr. Paul Müller, DDT's inventor, won the Nobel Prize for physiology and medicine. Ten years later, birdlovers were noticing that the chickadees, robins, and cardinals were disappearing, and Rachel Carson, in her dramatic warning, *Silent Spring,* spelled out for us the price we might have to pay for our enthusiastic, heedless use of pesticides.

This book describes safe, effective alternatives to chemical poisons. Using these less-drastic methods, you probably won't get instantaneous results, but have a little patience. If you keep working at these nontoxic ways, the intruders will either die or leave.

In many cases you'll have to use more than one approach at the same time—closing off wall cracks and other pest entries and hiding places, tightening your housekeeping and ways of food handling, using mechanical means to kill the nuisances that do get through.

You'll need to curb your craving for "magic" solutions to what are essentially long-term problems. There are none. I have the same cravings and must remind myself not to grab the nearest chemical when a roach or clothes moth shows up. Instead, we all need to develop the habit of working steadily at simple measures that knowledgeable professionals know will control unwanted wildlife.

Don't expect 100 percent success. There's no such thing in any form of pest control. What you can expect is a home with very few pests, if any, most of the time, and a safer, more wholesome environment for you and your family all of the time.

SOURCES

Blondell, Jerome. "Pesticide-Related Injuries Treated in U.S. Hospital Emergency Rooms." Washington, D.C.: Health Effects Branch, Environmental Protection Agency, 1981 Calendar Year Report.

"Highlights of the Findings of the *National Household Pesticide Usage Study, 1976–1977.*" Environmental Protection Agency. Undated memo.

Savage, Eldon P. *et al. National Study of Hospitalized Pesticide Poisonings, 1974–1976. Final Report.* Fort Collins, CO: Epidemiologic Pesticide Studies Center, Colorado State University; and the United States Environmental Protection Agency. 1980.

"Three Convicted of Falsifying Data at NALCO's IBT Unit." *Wall Street Journal,* Oct. 24, 1983.

PART ONE

NO PESTS, NO POISONS

"I would not recommend using any chemical pesticides in the home."

> Dr. Paul Cheo, former chief of research
> Los Angeles State and County Arboretum

On seeing a list of "safe" chemical pesticides, Corinne Ray, administrator of the Los Angeles Poison Information Center, shook her head. "They're all toxic," she said.

CHAPTER 1

Controls, Not Chemicals

Drawbacks and Limitations of Chemicals

Have you ever set off a bug bomb in your home and then felt sick for a couple of days? Or spread snail bait among your seedlings and prayed that the dog—or the kids—wouldn't sample it? If so, you've probably wondered, "Aren't there safer ways to fight pests?"

Modern pesticides usually bring quick death to those unpleasant little creatures determined to move in with us, but what do you do when the roaches or fleas or snails show up again and again, as they're very apt to do? Keep laying on the poisons?

A slow, searching walk along the pesticide aisles of a hardware store or garden shop can be a disturbing experience. From the acrid odors, whether from the insecticides or their dispersal agents, you know instinctively that these products can do you no good. Your instincts are right. A chemist once said, "If you can smell it, it will probably do some harm."

Scanning the package labels can make you even more uneasy. "Do not breathe spray mist. In case of eye or skin contact, flush with plenty of clear water. Contains a cholinesterase inhibitor" (a substance that impairs nerve function but the label doesn't tell you *that*).

3

"Bait may be attractive to dogs. Toxic to birds and other wildlife. Keep out of lakes, streams, and ponds." And always, "Keep out of children's reach." If these materials interact with air, you could be breathing toxic vapors as long as residues are left.

What's worse, chemicals aimed at a particular pest may be found unsafe by the Environmental Protection Agency yet still remain on market shelves. Confused consumers rely on retail sales help as ignorant as themselves.

Actually everyone is vulnerable to these toxins—the home gardener accidentally spraying a bug killer into the wind; people with allergies, respiratory ailments, or skin problems; the elderly; the very young; pets.

Pesticides Boomerang

Okay, so you're young, have neither children nor pets, and your skin and lungs are in great shape, free of allergy. You're also absolutely sure that you *always* handle hazardous substances without ever endangering yourself or anyone else. This fact, however, may give you pause: Some of our worst pests are now resistant to nearly all the chemical killers we have. Hundreds more are immune to at least some pesticides.

Among the species that readily bounce back from the chemicals are those most dangerous to us—flies, fleas, mosquitoes, ticks, and lice. Other resistant insects include major pests of agriculture, forests, and stored foods.

So chemicals may not solve your problem at all, even in the short run. What's more worrisome, homeowners who pour unused pesticides down the drain or drench garden soils with them are major polluters of our water. They're also contaminating our food. In September it was reported that shellfish taken from the coastal waters off Southern California contained dangerous levels of pesticides.

If you fall into the trap of using more and more pesticides, you run the risk of making yourself sick or developing an allergy to them. Meanwhile, the pests you're trying to kill, hardly fazed by the same substances endangering you, keep right on coming. If we add to these hazards the pollutants from automobiles and industry already infiltrating our bodies, doesn't it make good sense to keep our personal space as free of contaminants as possible?

There Are Other Ways to Fight Pests

Are there other ways to discourage and drive off the insects and rodents who find our homes and gardens attractive—ways that work for good? Yes, there are. Our ancestors knew how to get rid of all sorts of unwanted wildlife, inside and out, without poisons. With a little effort, two of their most powerful weapons are yours for the taking—cleanliness and sunlight. Other methods they used were often not much more complicated than these.

Insects and rodents have been our unwelcome companions ever since humans first put on clothes, moved into caves, and learned to grow and store food. In making life more secure and comfortable for ourselves, we also made it more secure and comfortable for many species of little wildlife.

Few of us want to move back out under the trees and run naked to get rid of the pests, and most of us would find a diet of roots and berries pretty dull. Down through the years, however, observant individuals noticed that insects and rodents avoid certain substances or conditions that humans find harmless or pleasant. They also noticed that lacking food or water, the little creatures move out.

Our ancestors, wherever they lived, stored food against the inevitable times of shortage. Some dug a hole and lined it with straw to keep insect raiders out of the stockpiles. The Chinese still build small shelters of clay and soft mortar, as their forebears did, to safeguard their grains. Early peoples also mixed their grains with aromatics like bay or eucalyptus leaves or inedibles like sand, ashes, or sulfur to repel pests. Farmers in Galicia, Spain, still build small granite and wood huts, called *horreos*; unchanged in design since about 500 B.C., these granaries are almost completely rodent-proof. In India, grains were once spread on a rug in the sun to drive out light-fleeing insects.

Many of the control methods described in this book are more modern versions of these techniques. They may have kept your grandmother's house pest free and given it the pleasant fragrance you remember. In today's enthusiasm for instant solutions, we tend to ignore safer, somewhat slower ways. Yet these less drastic methods make sense. After all, you wouldn't go after a gang of neighborhood burglars with a bomb when strong locks, good lighting, shrill alarms, and an alert police force can do the job effectively year after year.

Not all of the techniques described here are equally effective in all situations, but then neither are the chemicals. Climate, seasonal weather, natural surges in pest population, soil, your own thoroughness and persistence may all affect how well they work for you.

You, not the Pests, Are the Intruder

When that lovely home you enjoy was built, it displaced thousands, possibly millions, of tiny animals, all well integrated into an environment with adequate food and water as well as safe nesting sites. To the extent that your home provides these basics of life, insects and rodents will try to occupy it. Indeed, your home may actually be a better environment for them than the outdoors. Besides a good supply of dry food and water, most houses have innumerable nooks and crannies where small creatures can hide from predators, including humans.

Like all displaced individuals, your land's former occupants need a new home. Unless you're unusually hospitable, you don't want to share yours with the many-legged. So how can you get them out—and keep them out—of what's now your turf?

The following pages describe safe controls for most of the common pests that want to eat your food, clothing, furniture, plants, and even you. Many of these methods have been known for generations, and the new technology of Integrated Pest Management (IPM) has developed, and continues to develop, sound new ones.

Integrated Pest Management is a strategy (originally designed to help farmers) that uses technical information, ongoing monitoring of pest populations, crop assessment, and other techniques to control pests. Chemicals are just one part of the strategy, and their use is kept to a minimum. IPM is thought by many pest-control professionals to be the only way we can win our struggle against destructive insects and rodents without damaging ourselves, our domestic animals, or our environment.

Insects—Awesome Fertility, Phenomenal Resistance

First, however, you need to accept the fact that you're in a war you can't win. There are just too many insects.

Estimates of the number of *kinds* of insects on earth vary from .75 million to 1.5 million distinct species. Some experts think there may be as many as 10 million. Individual insects are beyond counting. Scientists can only make picturesque guesses like: The total weight of all insects is far greater than that of all other species combined. Or, the progeny of two flies mating in April would, if all survived, cover the earth by August with a disgusting blanket 47 feet thick.

Other insects have a reproductive ability just as awesome. The female of some species lays a million eggs in her lifetime, eggs that in many cases can hatch months or even years later. What's more unsettling is the fact that many pests become immune in a few generations to the most deadly insect poisons chemists have devised, and insect generations are often measured in weeks. (Rodents, which produce several litters a year, are also showing resistance to once-lethal substances.)

Today the housefly is hardly slowed by DDT, and one common pantry beetle has been found thriving on a diet of belladonna, aconite, and strychnine, all fatal to other life forms. Not surprisingly, no insect species has ever been exterminated. They are the most hardy and adaptable creatures on earth.

Clearly, this enemy is invincible. However, with vigilance and a sound strategy you can reach an acceptable stalemate. By using the controls described in this book, you will also lower your risk of accidental poisoning or illness.

Basic Antipest Strategy

Your basic strategy in getting rid of your small freeloaders is this: Find out where they're getting in, where they're finding food and water, and what conditions kill them or drive them off. Then launch your campaign by building them out, starving them out; rotating your food, clothing, and furniture; zapping them with very high or very low temperatures; exposing them to sunlight and fresh air; using fragrances they dislike; and trapping them.

Build Them Out

A tightly built, well-maintained home is essential in any pest-control campaign. An uncapped, unused chimney gives easy entry to rats, flies, spiders, and mite-infested birds. Defective roofing can have the same effect, while gutters and downspouts choked with dead leaves are ideal insect nurseries. Broken windowsills, leaky plumbing, and loose flashing around chimney and vents make for damp walls and the mold that many insects feed on.

Small cracks in walls, both inside and out, give roaches, ants, and rodents easy access to the interior of your home and safe hiding places

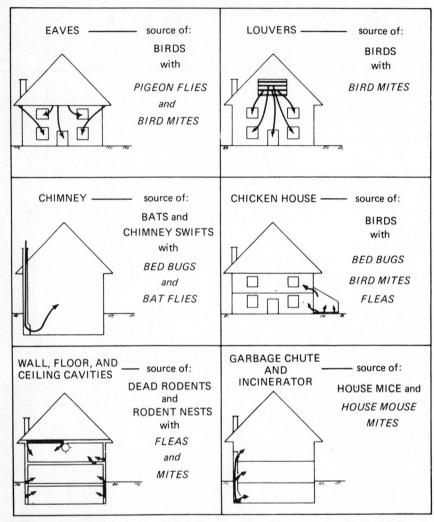

Figure 1.1. Common sources of bird, bat, and rodent ectoparasites annoying to humans. (*Household and Stored-Food Insects of Public Health Importance and Their Control.* Home Study Course 3013-G, Manual 9. Atlanta: Centers for Disease Control, 1982.)

once they're inside. Most pests need only a tiny fraction of an inch to establish themselves in a house. Damaged air grilles in attics and basements are another likely invasion route. Openings around plumbing pipes are the most frequent entry for cockroaches.

Inspect and Repair Your Home

Your first step, then, is a careful inspection of your dwelling. You may need professional help here: A qualified roofer and a sheetmetal contractor to make accurate assessments of the state of your roof, gutters, and flashings; a carpenter to point out needed repairs of windowsills and door frames. You yourself can take care of small cracks and holes in the outside walls with a caulking gun, latex paint, or patching stucco.

Tight window screens in good repair and screen doors that close automatically and quickly are your best defenses against flying pests.

As any slum dweller can tell you, broken toilets and rusted-out floor drains are a common way rats get into a run-down building. So make sure that all your toilets are tightly bolted to the floor and work properly. The strainers covering drains in laundry tubs, showers, bathtubs, and basement or garage floors should be strong enough and of fine enough gauge so that no rat or cockroach (both sewer dwellers) can push its way through them.

Modify Landscaping

Look at your landscaping with a skeptical eye. Though they make your home attractive, the shrubs and flowers hugging its outside walls can be the staging area for an invasion of mites, pantry weevils, or spiders. A clear strip at least eighteen inches wide, either of concrete or sand, circling the foundation will act as a barricade against most crawlers. If you have Algerian ivy bordering your lawn, pull it up. Its broad leaves, growing close to the ground, are a perfect habitat for black widow spiders, rats, and mice.

Dampness Draws Many Insects

Your home may be tight as a drum yet still attract insect pests. Interior dampness is probably the cause, and it can come from several sources. In a structure under 30 years old, your problem may be built in. With the hollow-wall construction now common throughout the United States, moisture either from air-conditioning or central heating, condenses in the wall voids. In time, fungi develop there, to be followed by mold-eating insects. From these secluded feeding grounds they can easily find their way into your living quarters through vents, electrical outlets, and wall switches.

In addition, today's rooms tend to be small and relatively airtight, with less air to absorb and disperse the moisture. Many homes are built on slabs and lack subfloor ventilation. The slab itself, made of cement and water, gives off moisture long after it's been poured. And sliding glass doors admit garden crawlers—millipedes, centipedes, sowbugs. Enclosed patios, lush with foliage in wooden or brick planters, also draw moisture-loving insects. The planters themselves, unless waterproofed on all sides, sheathed from the soil below, and set against a crack-free wall, are potential gateways for termites. Sprinkler-drenched stucco and damp crawl space also favor subterranean termites.

Our passion for washing compounds the problem. Dishwashing, bathing, and laundering for a family of four add about 3 gallons of water to their immediate environment every day, a total easily tripled by a high water table and lawn watering.

If you think your home is extremely damp, you should consider buying a dehumidifier as a major pest-control step. In less severe conditions, an air-conditioner also dehumidifies inside air.

Starve Them Out and Clean Them Out

Now that your home is as pest-tight and dry as possible, you can begin to starve out those raiders who were there before you began your campaign or who managed to slip by your barriers, as some occasionally will. Cleanliness and proper food storage are your strongest weapons here. Regular vacuuming of carpets, upholstered furniture, draperies, and shelves sucks up adults, larvae, and eggs along with any food crumbs that could sustain them.

If automatic garbage disposals are not used in your community and the garbage collection service is inadequate, your trash barrels are probably drawing and incubating flies. Become a political activist; join with your neighbors to fight city hall and demand more frequent pickup. Until service improves, wrap food garbage in plastic bags and close them tightly before putting them outside. Deodorize your barrels with a weekly rinsing with borax solution.

How about the area around your washer and dryer? Are there lint fluffs near the dryer, damp spots under tub and washer, bits of soap here and there? Warm, sheltered dryers are favorite nesting sites for rodents, who use the lint to keep their young cozy. Tiny amounts of water satisfy thirsty roaches, who also find soap a satisfying meal. So, vacuum up the lint (be sure to move the dryer away from the wall), clean up the soap scraps, and make it a point to keep the area dry.

Spring Cleaning—Still a Good Idea

Do you remember your mother's or grandmother's annual flurry of spring cleaning? Her main reason for all the dusting, rug beating, and corner scouring may have been to freshen the house after a winter of stuffy heated air. But whether she knew it or not, she was also disrupting the reproductive cycles of house pests who'd laid their eggs the summer and fall before. Such interruptions seem to slow the biological clock of developing insects.

She sunned the rugs for hours on a washline and then thrashed them with a bamboo beater. Then she took down the heavy draperies and brushed out all the folds (safe havens for tiny creatures) before storing them in an airtight chest for the summer. She emptied dresser drawers and dusted them, cleaned kitchen shelves and changed their paper lining. All the bustle left the home fresh and airy, pleasantly tidy. More important, her intense housekeeping activities were excellent pest preventives.

This annual rite of spring seems to have faded away as more and more women work outside the home, but whatever the season, it would boot out your unwanted guests.

If the freshening-up sets you to thinking about redecorating, consider taking up your wall-to-wall carpeting—a good hangout for fleas, carpet beetles, roaches, and their ilk—and replacing it with area rugs. They're easy to lift for thorough cleaning, and a waxed wood or vinyl floor around them offers little comfort to a tiny intruder looking for a hideaway.

Proper Food Storage

How do you store your staples? Keeping dry foods in bags and cardboard boxes is like displaying a menu for grain and legume pests. Tightly closed metal, glass or plastic containers lock these bugs out, while separating new stocks of food from the old prevents the spread of any possible infestation. Store anything long enough and something will come along and eat it, so buy your groceries in amounts you're apt to use up in a short time.

Clear Out the Clutter

A cluttered home attracts insects and other small creatures. Its undisturbed corners offer them innumerable safe havens so, of course, they'll move in. Soon their relatives show up. If you're an incurable "saver"— always keeping string, boxes, old newspapers, bits of this and that against some vague need down the road—try to break the habit. Paper and cardboard, long-forgotten stuffed toys, and clothing unworn for years all eventually draw wildlife, so throw out the junk and give what's still usable to the Salvation Army.

Incidentally, while cleaning all those neglected places, be sure to wear sturdy work gloves in case you meet up with a spider resentful of your intrusion.

Even after your house is no longer attractive to the most obnoxious pests, airing and sunning rooms frequently makes them feel unwelcome.

Shake Them Up

Rotating your clothing so that all garments are worn and then cleaned or aired frequently is one of your best protection against moths, silverfish, and other fabric fanciers. Fashion experts advise, "If you haven't worn a particular garment in two years, give it away." That's good advice whether you want to be stylish or just free of moth holes.

While wearing woolens, for example, you disturb any moth larvae lurking in the folds, brush them or shake them off the fabric, shrivel them when you walk in the sun, or drown them when you launder the article. Dry cleaning also kills them. Who needs an additional chemical?

Do you enjoy rearranging the furniture? You may do it to be creative, but you're also making life unpleasant for resident insects. Exposing carpets and floors under heavy furniture makes it easy to vacuum up carpet beetles and other undesirables. But if you're like me and don't ever rearrange furniture once the movers leave, you're asking for trouble.

Rugs and carpets under low-slung, heavy chests, sofas, and beds are hard to reach with most vacuum cleaners, so moth larvae, carpet beetles, and other pests of textiles can gather there. Don't forget to rearrange the pictures on the walls while you're at it because bedbugs and roaches can set up housekeeping behind a frame and under its paper backing.

Roast Them Out or Freeze Them Out

Long before anyone dreamed up chemical insect killers, it was known that very high and very low temperatures eliminate most household pests. Few survive prolonged exposure to subzero conditions or heat of 140 degrees Fahrenheit. For generations, cold storage has been the standard summer protection for furs vulnerable to moth damage as well to preventing drying of the skins. Before World War II, European exterminators regularly blew air heated to 152° Fahrenheit into apartments to clean out infestations. With your furnace or your freezer, you can apply the same principles.

Trap Them, Barricade Them, Repel Them

If you know what a specific pest likes to eat, you can set up a trap that is harmless to humans and pets but lethal to the insect you're struggling with. For example, a simple trap baited with meat or syrup can kill hundreds of flies a week; a small jar half filled with beer finishes off cockroaches by the dozen; and silverfish can be done in with a trap partly filled with flour.

If you'd rather keep a pest away, try a simple barricade. For example, a smoothly finished removable metal grid level with the pavement outside your doorway will stop any crawlers headed inside. A snail or slug can't inch past sawdust. Caterpillars can't negotiate a band of hay girdling a tree. Vaseline stops ants dead in their tracks. By thinking along the same lines, you could come up with solutions as simple and effective as these.

You might also want to explore herbs to repel unwanted wildlife. Insects have keen senses of smell and avoid sharp scents like mint, tansy, basil, cedar, and camphor.

Garden Controls

Controlling garden pests needn't take any more effort than fighting those in your house. Obviously you can't build pests out of a garden, but some of the techniques that work indoors, such as good sanitation and mechanical barriers, can be effective outside.

You can also apply other strategies to a garden. Avoid planting large stretches of one plant or shrub. This practice, called monoculture, spreads a banquet for insects that feed on that particular vegetation, and is the reason farmers have to depend on chemical pesticides. You might consider planting an English-type garden instead, with colorful profusions of different flowers and shrubs. Timing the planting and harvesting of vegetables so as to avoid their insect pests is another way to outwit the freeloaders.

Your best aids outdoors are your wildlife allies—birds, lizards, bats, toads, predatory insects—who feed on all insect stages. One of the biggest drawbacks of chemical pesticides is that they kill friend and foe alike; and when the predators are gone, the pests return in greater numbers than before since you've killed off nature's most powerful controllers. Making matters worse, insects that were formerly no problem at all, suddenly free of their enemies, become major headaches. One of the clearest examples of this is the surge in spider mites, once minor troublemakers, to the status of a worldwide threat to forests and agriculture.

Patience and a Little Tolerance

If you tend to panic and grab automatically for a spray or fogger when a bug comes near, try to get that knee-jerk reaction under control. The chemical could do you more harm than the insect you're trying to kill. That little animal may not even be a pest at all.

In fact, only about .25 percent of all insect species are pests. Some scientists believe that without the arthropods (as insects, spiders, and mites are called), the human species would die out. Most species are beneficial. They turn decaying matter into plant food. By helping plants reproduce, they feed us, keep our earth green and beautiful, and cleanse our air. In a never-ending warfare among their kind, a great many of them destroy the insects who would destroy our food or endanger us.

And who would want to kill a graceful swallowtail butterfly or soaring dragonfly (which, incidentally, preys on mosquitoes)?

When you find roaches touring your kitchen sink or hear the patter of rodents' paws in your walls or attic, ask yourself: How did they get in? What are they finding to eat and drink? Where can they hide? What are the safest ways to get them out? The steps you take in answering these questions will get matters under control.

Ultimately you may have to apply a chemical—but not nearly so much or so often as you would have if you hadn't first made your home less attractive to the invaders. Indeed, with reasonable care, you may never have to use a chemical pesticide again.

CHAPTER 2

Is Your Problem Really Insect Pests?

In her dramatic book, *Silent Spring,* warning of the dangers in our heavy reliance on chemical pesticides, Rachel Carson told the story of a woman deathly afraid of spiders. Finding a few in her cellar one August, she sprayed the whole area thoroughly—under the stairs, in the storage cupboards, around the rafters—with a solution of DDT and petroleum distillates. Wanting to make sure she'd killed every last weaver, she repeated the spraying twice more at intervals of a few weeks. After each go-around, she felt nauseated and extremely nervous.

Following the third spraying more alarming symptoms appeared: phlebitis, fever, painful joints. Shortly after, she developed leukemia and died.

Although the culprit may have been the petroleum distillates and not the DDT, the tragic irony of her suffering and death is that the spiders were keeping her cellar free of pests. They may even have been relatively immune to the substance she was spraying. Furthermore, she could have cleaned them out more effectively—certainly more safely—with her vacuum sweeper.

Fear of Insects

This woman was a victim of entomophobia, the unreasoning fear of insects and like creatures. While hers was an extreme case, revulsion caused by insects and other tiny animals is widespread. Many of us pull back in fear when an odd-looking beetle crosses our path. We know little if anything about it, yet we are repelled. Maybe it eats things we value, such as our garden plants or our food. More likely it actually helps us by serving as food for animals useful to us, like birds and lizards, or by preying on even smaller beings—aphids and scales—that destroy plants. In addition, it may aerate the soil, letting us grow healthier plants.

Most of us can't predict or control an insect's behavior. Unlike most other wild species, it doesn't seem to be afraid of us. Indeed, it ignores us unless we're about to crush it with our feet. Is that what seems so threatening—that we have almost no impact on another living being?

Yet, although insects can get along very well without us, we cannot ignore them. They were around hundreds of millions of years before the first humans showed up and, from all indications, will probably still be here long after we've vanished. But while we're here together, we depend on them for our food and a livable environment.

Not everyone who's afraid of insects carries it to the point of slow suicide, but educated individuals can work pretty hard at it. A retired schoolteacher friend recently announced that she'd just installed an electronic bug zapper in her garden, already thoroughly doused with chemicals. When I told her she was killing insects that would hold down the pests, she grimaced in disgust. "I hate bugs," she said, "*all* of them." One wonders how many children she infected with her fear.

Actually, her gadget attracts more winged insects than would probably come to her garden without it. Meanwhile, an evening's conversation on her patio is punctuated by the crackle of raw electricity as a living creature is electrocuted every few seconds.

Psychologists aren't sure what causes phobias. According to the *New Columbia Encyclopedia,* they arise from inner conflicts possibly rooted in childhood insecurities. Besides fear of insects, other phobias include fear of heights; enclosed spaces; birds; cats; even other people. The entomophobic person's constant spraying for insects, harmless or annoying, is a kind of defense mechanism to overcome anxieties caused by deep-seated emotional distress.

Overcoming Fear of Insects

A person with a dangerously intense fear of insects or other small animals can overcome these rampant feelings. Suppose, for example, that you're terrified of spiders. The following steps can gradually build up your tolerance:

- Start by looking at pictures of spiders.
- Read about them.
- Get up early and watch a garden spider spin a web, one of nature's most exquisite structures.
- Stay in a room with a spider for a little while.
- Examine it with a magnifying lens.
- Let one walk over your hand (now give yourself a medal!).

Should all these be of no help, find a good psychological counselor because such extreme fear can cause you physical harm.

Advertisers play on this widespread, irrational fear of the insect world. Nightly in 1983, American television viewers watched a middle-aged couple quiver in terror because he's found "them" in the bathroom. After setting off an aerosol chemical, they collapsed in relief. "They" were gone. What the ad failed to say was that unless the neurotic pair change whatever is making their bathroom attractive to "them" (most likely a leaky plumbing fixture), the insects will *always* return.

Another series of commercials had Muhammed Ali urging us to get the "b-u-u-gs" under control by fogging the air with a chemical. Cartoons of little crawlers marching into the mist and then obligingly bellying up emphasized the message.

What the helpful heavyweight didn't tell viewers was that there are other, longer-lasting ways of control. The insects he was talking about looked like cockroaches, and the champ failed to say that cockroaches learn to skirt most chemical poisons.

Children and Other Little Creatures

Children usually have no such hang-ups. The younger they are, the more fascinated they seem to be with all kinds of small wildlife. At age 8, Susie loved to catch honeybees in a milk bottle filled with purple clover and bring the buzzing bottle to friends as a gift. Six-year-old Mike would grin delightedly when a monarch butterfly, newly emerged from its cocoon, rested on his shoulder.

Paul at age 7 loved to keep "Hamsterdam," his pet (what else?) hamster, nestled in his shirt pocket, its two black eyes an extra pair of shiny shirt buttons. Five-year-old Sarah would spend hours combing through her family's garden for pillbugs. When she'd find one, she'd lift it gently to her palm, where it would curl up in self-defense. "Ooh, look!" she'd whisper, "It's going to sleep," and set it tenderly back in the grass. Some 20 years later she regularly sets off bug bombs in her apartment.

Do we adults infect children with our own half-understood fears? Or is it that highly urbanized as we've become, we fear the outdoors and its wildlife?

Home gardeners, perhaps uncomfortable with nature uncontrolled, apply a whole pharmacy of pesticides to their plants. Wiping out the leaf nibblers, they also destroy the insects' natural predators, and so are trapped in a chemical treadmill.

If intense dislike of insects is your problem, the procedures suggested in the chapters that follow, by holding down the number that you encounter, can help you. But there are two problems many people struggle with that are beyond the scope of this work: imaginary insect infestations and "cable mite" dermatitis.

Imaginary Pests

Some people are convinced that their bodies are riddled with invisible insects that cause them great discomfort. So real do these nonexistent creatures seem to the sufferers that they often persuade their relatives that the infestation is real.

"We get half a dozen calls or letters a month," says James M. Stewart of the National Centers for Disease Control, "from people who are sure they're infested with one insect or another. One man thought that moth larvae were dropping on him and burrowing under his skin. We try to tell them what their problem may be without actually telling them that they're crazy."

To prove the existence of their tiny tormentors, patients will bring to a dermatologist or entomologist pieces of the animals' "bodies." These usually turn out to be skin scrapings, lint from their clothing, or products of the person's oil glands. One deluded man, a successful playwright, placed his "vermin" on a piece of paper and made it jump. He ignored the fact that he'd created static electricity by walking over carpet, said W. G. Waldron, the entomologist, who tried to help him.

The problem is a serious one, even life-threatening, although mistakenly identified. Researchers J. W. Wilson and H. E. Miller tell of a woman who boiled her family's clothing every night in an effort to dislodge "mites." She also made her children wash themselves with gasoline, turning them into walking Molotov cocktails. Hugo Hartnack, an entomologist in the 1930s, once tried to help a husky police officer who broke down and wept when describing his "infestation." Although the naturalist listened sympathetically, he could not help the man. A few days later, the police officer committed suicide.

If someone close to you thinks he or she has mites even though a dermatologist has said there are none, try to persuade the person to see a psychiatrist because this is a psychological problem, not a physical one.

Cable Mite Dermatitis

A spinoff of this delusion about insects is "cable mite dermatitis." People working in areas of low humidity and under considerable pressure sometimes develop a mysterious itch. They may be convinced that invisible organisms on the papers or telephones they're handling are causing the rash.

The power of suggestion is strong among groups of people working together, and the cable mite problem can be a serious one for a company. In September 1980, Republic Airlines had to send eighteen of the twenty employees in its Los Angeles office home sick in one week. They weren't sick from the dermatitis some of them had but from the pesticide used to wipe out the *nonexistent bugs*.

In a later incident, described by researchers H. G. Scott and J. M. Clinton, workers in a physics laboratory developed dermatitis, and were positive that their rashes came from some microscopic animals in the lab. Luckily, before the foggers and sprays arrived, someone discovered that rock-wool insulation in the ceiling was drifting into the ventilation system and causing the dry, itchy skin. In this case, some careful detective work and a slight modification to the building was the solution to a real dermatitis with an imaginary cause.

PART TWO

PESTS OF FOOD

---------------- CHAPTER 3 ----------------

Outsmarting the Cagey Cockroach

In the seventeenth century, any Danish sailor who caught 1000 cockroaches aboard ship received a bottle of brandy. A hundred years later, Captain Bligh ordered the *Bounty*'s decks drenched with boiling water to kill the voracious insects devouring his cargo of breadfruit trees. In the nineteenth century, for every 300 seafaring roaches he trapped, a Japanese tar won a day's shore leave.

The United States Navy sprayed 10,000 gallons of pesticide on its ships in 1978 in a futile effort to dislodge this pest. More recently, the Americans have tried sex lures as roach traps. After a 3-month test aboard a ship of the Atlantic fleet, the Navy admitted failure and gave up.

If powerful, efficient navies can't control cockroaches, what can the average householder do against these pests?

Plenty!

While not as exciting as shore leave, brandy, or sex, you can be rewarded with reliable pest control by tight maintenance of your home, cleanliness, and the strategic use of boric acid and traps. First, distasteful though it may be, you'll need to learn a few facts about the pest.

Nature of the Cockroach

These are the kinds of cockroaches you're most apt to meet up with (see Figure 3.1):

American. One of the largest: between 35 and 40 millimeters long. Reddish or dark brown. Prefers warmth. Eggs need high humidity. Often found in sewers. May be attracted to outside lights at night.

Brownbanded. Brown and yellowish bands across wings; between 10 and 15 millimeters long. Prefers warm, drier environment. Often found throughout house. Lays eggs on dark, vertical surfaces such as bookcases, draperies, closet walls. Often found high up in rooms. Can reach huge numbers.

German. The most common species. Grayish color; between 10 and 15 millimeters long. Two black bars on head. Prefers warm, moist areas such as kitchens, bathrooms; also garbage containers, shrubbery, crawl spaces. Extremely fertile; one female can have 30,000 descendants in one year.

Oriental. Blackish color, between 20 and 27 millimeters long. Smells bad. Drawn to excrement. Prefers damp, cooler areas such as basements, dense vegetation, water meter vaults, floor drains; also sinks, refrigerators, washing machines, water heaters.

What Cockroaches Eat

Roaches are primarily carbohydrate eaters, that is, they feed on just about anything of vegetable origin. Since they also like meat, this makes them omnivorous. They prefer starches but thrive on grease, sweets, paper, soap, cardboard, book bindings, ink, shoe polish, even dirty clothes. They've been known to gnaw the fingernails of sleeping sailors and infants' eyelashes. They're especially fond of beer.

True survivors, they can live 3 months without food and 30 days without water. They react faster than the blink of an eye to a footstep. Since they taste their food before eating it and learn to avoid chemically treated surfaces, most chemical pesticides do not give good long-term control.

Cockroaches Are Useful?

Despite their well-deserved bad reputation, cockroaches are of some use to us. Researchers have used them in investigating cancer, heart

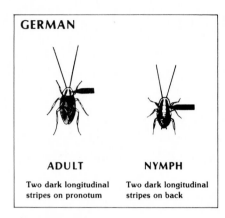

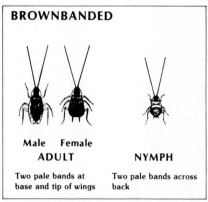

GERMAN

ADULT — Two dark longitudinal stripes on pronotum

NYMPH — Two dark longitudinal stripes on back

BROWNBANDED

Male Female
ADULT — Two pale bands at base and tip of wings

NYMPH — Two pale bands across back

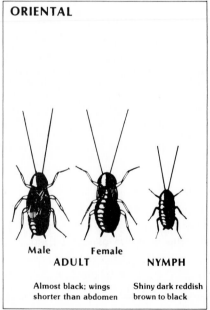

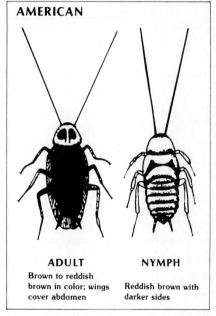

ORIENTAL

Male Female
ADULT — Almost black; wings shorter than abdomen

NYMPH — Shiny dark reddish brown to black

AMERICAN

ADULT — Brown to reddish brown in color; wings cover abdomen

NYMPH — Reddish brown with darker sides

Figure 3.1. If your Croton bugs, water bugs, shad bugs, palmetto bugs, black beetles, or Bombay canaries look like any of these insects, you have cockroaches. Most need a warm, dark, moist environment. (*Controlling Household Cockroaches*. Division of Agricultural Sciences, University of California, Leaflet 21035, 1978.)

disease, nutrition, and the effects of space travel. They're also vital to our forests, where they turn dead vegetation into valuable soil nutrients.

Exterminator's Bread and Butter

Chances are that you're reading this book because you've been sharing quarters with cockroaches. So common are these pests and so anxious are we to get rid of them that they've been called the exterminator's bread and butter. Few insects are so clever at making themselves comfortable in our homes. They've been found nesting in telephones, televisions, radios, refrigerators, electric clocks, even a flute. People have moved out of their apartments because they couldn't cope with a cockroach infestation. If you see one in the daytime, you can be certain that you're hosting many more. Roaches hide from light, and only expose themselves to it if they are crowded out of their hiding places.

Roach-borne Diseases

The cockroach is as fastidious as a housecat, forever cleaning itself and polishing its sleek shell, but if one should crash your next dinner party, you could find your circle of friends abruptly shrunken. Few creatures arouse such loathing—and with good reason. They are known carriers of:

boils	plague
dysentery	polio
hepatitis	salmonella
parasitic toxoplasmosis	typhus

Cockroach organisms of salmonella, the common cause of food poisoning, have survived on cornflakes and crackers for four years. Parasitic toxoplasmosis, widespread though unfamiliar, is harmless to human adults but devastating to an unborn child. If you find one of these vermin in any of your food stocks, throw that food out immediately.

Like all insects, cockroaches are extremely prolific but, unlike more seasonal species, they breed all year long, some even without mating. In recent tests 5000 roaches set loose in Raid's corporate kitchen had multiplied to 16,298 when the room was fumigated 1 week later. While in their eggs, which are clustered in yellowish purse-like cases, cockroaches are immune to all insecticides (see Figure 3.2). To put it mildly, this is a tough adversary.

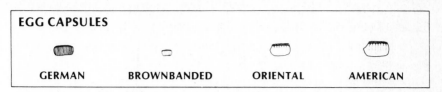

Figure 3.2. Cockroach egg capsules. (*Controlling Household Cockroaches*. Division of Agricultural Sciences, University of California, Leaflet 21035, 1978.)

Shut Them Out

Ridding Your Home of Cockroaches

The critical step in getting rid of these vermin is to change whatever makes your home attractive to them. Thorough cleaning of all infested rooms, and even those that seem to be roach-free, is a good start. Use soap and water on all washable surfaces and your vacuum sweeper on the others.

The Kitchen

- clean the area around your water heater, especially if it's housed in a separate closet.
- move the refrigerator and stove out into the room and scrub the wall and floor behind them with an industrial strength cleaner.
- when the floor is dry, dust it lightly with technical boric acid powder.
- if wallpaper behind your appliances has come loose, glue it down securely. Loosened paper makes a fine nesting site for insects.
- while the refrigerator's away from the wall, check the outside of the box for threadlike brown droppings. These, along with an unpleasant odor, are a sure sign of roaches in the insulation, a favorite and hard-to-reach harborage. Thousands can live in one refrigerator. If you find roach droppings, have the box fumigated—or get a new one.

If you're shopping for a used refrigerator, "beware" is the word if the seller is a private party. You could be buying a cockroach infestation along with that bargain box. You'd be better off to find a reputable dealer in used appliances, who will have had all the refrigerators fumigated as a matter of course.

A look under your kitchen sink could tell you why you're having a problem. Do you keep bags or boxes of pet food there, along with your damp dishrag? If so, you're duplicating a roach breeder maintained at one time by the famed San Diego Zoo to raise food for its exotic birds. The zoo's birdkeepers housed large colonies of the insects in 20-gallon garbage cans filled with rolls of corrugated cardboard surrounding a jar of water. In the jar was a cotton wick. The whole assemblage was generously sprinkled with dog kibble. So, dry your dishrag in the open and store your pet's food in a tightly covered can.

The Bathroom

Here, water rather than dirt is apt to be the roach lure.

- Danger spots are wherever moisture gathers—the base of the toilet, the angle between countertop and backsplash. Dry these up and keep them dry.
- If you have a vanity cabinet, check for water around the pipes underneath.
- After using your rubber shower mat, hang it up to dry. Its rows of suction cups can give a roach a custom-fitted hiding place.

Other Places

Although this insect prefers kitchens and bathrooms, it easily fans out to other parts of the house.

- Closets, TVs and radios, electric clocks should all be inspected and cleaned thoroughly.
- Vacuum any upholstery near the TV to pick up crumbs from snacks eaten there. Confine the family's snacking to the kitchen or dining room.
- Empty bookcases and vacuum the shelves. If they're washable, scrub them. Cockroaches like the binding glue in books, and one species attaches its egg cases to dark vertical surfaces.
- Shake each book out and flip its pages wide open. A couple of cockroaches along with some silverfish may tumble out.
- Vacuum the folds in your window draperies. These are also dark vertical surfaces.

- Vacuum or wash corners under tables and chairs, favorite hiding places for brownbanded cockroaches.

- Dispose of the vacuumings by burning them, burying them deep in the ground, or placing them in a tightly closed garbage can set in the hot sun. Any of these methods kill the vermin's young.

- Check your garage, basement, and attic and get rid of those piles of old magazines and newspapers, boxes, and old clothes. All can support large colonies of roaches. Corrugated boxes are especially bad, because the corrugations easily shelter the tiny cockroach nymphs.

Small, Important Repairs

Now's the time for long-delayed small repair jobs. That dripping faucet and seeping toilet are a reservoir for insects. A new washer may take care of the faucet but the toilet may need a plumber's services.

Fill all cracks in the walls and woodwork, including the crevice between baseboard and floor, any spaces between cupboard walls and shelves, cracks around drainpipes and sinks. A caulking gun does a good job. Where the cracks are hair-fine, latex paint can seal them off. Before deciding to ignore a crack, remember that a roach can hide in a space $\frac{1}{16}$ inch high. If you're short of time, stuffing these hideouts with steel wool is a good stopgap.

Has your sink's splashboard or its counter come loose? If so, glue them down securely. Both are often the main sites of an infestation. Be sure to caulk all joints.

Outdoors: Repairs and Housekeeping Important Here Too

After the interior is pest-tight, set to work on exterior stucco or cement with the right patching compound to close up even the smallest opening. Minimize dampness wherever it is.

- Seal any gaps around water or gas pipes coming into the house. These are virtually cockroach expressways.

- Put several inches of gravel into your water meter vault. Its damp cavity is a perfect roach incubator.

- Pull up dense patches of ground cover, especially Algerian ivy. Thin out other plants to eliminate dampness.
- Clean up debris.
- Move stacks of lumber and cordwood away from the house.

Apartment Dwellers

Does your apartment sparkle, yet you still spot occasional scurrying brown crawlers when you turn on the kitchen or bathroom light? The insects could be migrating into your place from the neighbors'. To stop the intruders, nail window screening over the heating vents between your quarters and other tenants', and weatherstrip your door to the common hall.

Be especially watchful if your building has an incinerator or a compactor. When these don't function properly, or the area around them isn't kept clean, they attract vermin (see Chapter 6 for effective incinerator temperatures).

Some Useful "Don'ts"

- Don't save grocery bags. They're frequently infested, and even if not, they are harborage for roaches that come in by other routes.
- Don't buy beverage cartons with spilled syrup or malt. You could be buying roaches. One soft-drink bottle returned for deposit held 200 young vermin.
- Don't save corrugated cardboard boxes. Roaches love them. If you're collecting boxes before moving, look them over to make sure you don't carry a verminous family along with the pots and pans to your new home.
- Don't buy used furniture without inspecting each piece with a gimlet eye. That mellowed dresser or bed could have some unwanted occupants.

Starve Them Out

Managing Food, Garbage, Water

Cockroaches in your kitchen are a loud signal to change the ways you manage food and water. These questions and answers tell you how.

- Are your staples in tightly closed, impenetrable containers? Clean pickle jars, tea canisters, tin cracker boxes are all suitable for storing dry foods; you don't need decorated kitchenware for this job.

- Do you always clean up spills and crumbs promptly and sweep the kitchen floor daily? From an insect's-eye view, a few crumbs are a feast.

- To prevent pest-drawing odor during the day, do you accumulate garbage in a plastic bag, which is then closed with a twist tie?

- Do you put your garbage out every night? That's when roaches are most active.

- Do you leave snacks out for your pet? Kibble is as delicious to vermin as it is to dogs and cats. Remove the animal's food right after feeding time.

- Do you leave your pet's water out overnight. If the animal insists on a nighttime drink, set the water dish in a larger pan filled with detergent suds.

- When do you wash your dinner dishes? If you absolutely *must* leave them unwashed for a few hours (a practice definitely not recommended), submerge them in water and detergent.

- Are there little reservoirs around your home? The most unexpected watering holes attract these insects. So cover fish tanks, flower vases, and not-quite-empty soft-drink bottles; refrigerate overripe fruits and vegetables and dry up the catch pan under your refrigerator.

Wipe Them Out

Boric Acid: The Proven Killer

Now that your home is no longer fit for cockroaches and you're not encouraging them outdoors, you can eliminate those left inside. Before grabbing for the fogger or spray, remember that after repeated use, chemical pesticides don't affect this insect. For the long haul the safest, most potent roach killer is technical boric acid.

Known for generations as a reliable medical antiseptic, boric acid powder seems to work as an insecticide in two ways: While preening itself, the animal swallows the powder it has picked up on its feet. Boric acid also penetrates the roach's outer covering.

Although the insect has learned to avoid chemical pesticides that it can smell and has developed immunity to others, it has never learned to avoid boric acid or become resistant to it. Boric acid has been used successfully against roaches since the 1940s.

Use Only Technical Boric Acid

It's illegal as well as unwise to use medicinal boric acid against roaches. It's too easily confused with table sugar or salt, and a tablespoonful swallowed can kill a small child and give an adult digestive upset. Technical boric acid is tinted blue for easy identification, and is treated electrostatically to cling to the insect's coat. The powdered roach carries the poison back to its lair, where others pick it up.

Sprinkled in out-of-the-way corners, technical boric acid does no harm and since it does not interact with air, continues to kill any roach it contacts for as long as it's in place. It's available in hardware and builders' supply stores.

Amount Needed

About 1 to 2 pounds will take care of an apartment; 2 to 4 pounds will clear out a moderate-sized home.

How to Apply Boric Acid

The best way to use it is to place small amounts in secluded areas so as to form a light layer on all surfaces. Heavier layers can harden if water reaches them, destroying their effectiveness.

Apply the powder along kitchen and bathroom baseboards, behind and under stoves and refrigerators, under your kitchen sink, at the back of the bathroom vanity cabinets, behind toilets. (See Figure 3.3.) You can also blow it into the razor blade slots of the medicine cabinet to disperse it in the wall void behind the bathroom.

Do not sprinkle it where you store your food. To penetrate the empty spaces under your kitchen cabinets, a great cockroach lair, drill a ½-inch hole every 5 or 6 inches at the top of the kick panel or in the bottom shelf of each floor cabinet and blow in the boric acid with a Getz powder blower, a bulb duster, or a squeeze bottle such as is used for ketchup or mustard. *Be sure to label the container clearly and permanently.* Electrostatically treated boric acid does not lump.

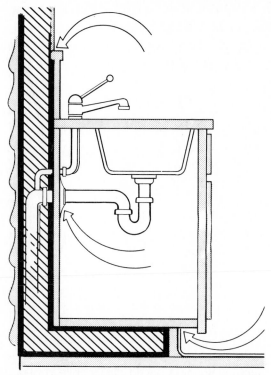

Figure 3.3. Application of boric acid powder around kitchen sink. Dark lines indicate a thin layer of the powder. Access to these areas may be found at corners through small cracks and at plumbing and electrical openings in the wall. (*Controlling Household Cockroaches*. Division of Agricultural Sciences, University of California, Leaflet 21035, 1978.)

At night look for other hiding places. Enter your darkened kitchen and bathroom and shine a narrow-beamed flashlight on likely lairs. The roaches will run from it, but not before you've had a chance to see where they're going. Then, treat those areas.

Boric Acid: Slow but Sure

Boric acid works more slowly than other antiroach products. A day or two after applying it, you'll notice one or two small corpses; or you may see more of the live pests than you have before as the dying insects wander out of the walls. Satisfactory control takes anywhere from 2 to 10 days, by which time the number you see will have dropped sharply. If you keep up your good housekeeping and home repairs, months or years may go by before anymore show up. The fatal powder just keeps on working.

A Note of Caution: Because boric acid is very slippery, don't sprinkle it where people walk. If you're going to be in contact with it for an extended period while placing it around the house, wear gloves and a particle mask. Also, plants are easily damaged by the powder, whether by direct contact or through the soil. Keep it away from children and pets.

An ideal time to apply boric acid is while your house is being built. It's easy then to dust attic, wall voids, dropped ceilings, and voids under cabinets and built-in appliances.

Silica Gel

Another powder that clears out cockroaches is silica gel sorptive dust. Applied like boric acid, it absorbs the insect's waxy coat, and the animal shrivels and dies. Silica gel, however, is very repellent and the pest may soon learn to avoid it. Besides costing more than boric acid, silica gel is highly irritating to the lungs and tends to float all over the house, clinging to curtains and absorbing wax from furniture.

Roaches from Sewers

If you've been finding cockroaches in the sinks in the morning, they may be climbing up from the sewers. *Don't pour a pesticide down the drain.* It's against the law in many communities and harms the environment everywhere. The chemical can also deactivate a septic tank.

Call the local sanitation department or public health authorities to come out and treat your sewer lines. Until they do, close all your drains at night and stuff the overflow hole with a wad of steel wool wrapped in a small plastic bag.

Traps

If you have very young children, here are some traps that have proven effective.

The British Trap

1. Wrap masking tape (to give your prey a foothold) around the outside of an empty jam jar.
2. Half fill jar with a mixture of beer, a few banana slices or pieces of the peel, and a drop or two of anise extract. Boiled raisins are also good baits.
3. Smear a thin band of petroleum jelly 1 or 2 inches wide just inside the rim so your victims can't climb out.

Twelve such fragrant lures are reported to have netted over 8000 vermin in the United Kingdom during a 3-month period.

The Texas Trap

1. Paint a wide-mouthed pint jar flat black.
2. Smear petroleum jelly inside the rim as above.
3. Add pet kibble or pieces of apple, potato, or banana peel.

Whichever you choose, set a few of them upright against corners of the room or under a sink, wherever you've seen roaches. To catch German cockroaches, the most common, traps must be inside; those placed outside will catch other species. (See Figure 3.4 for trap placement.) When these traps start working, leave a few victims in them to lure others. Drown your catch in a bucket of hot, sudsy water. Of course, always wash your hands after handling the traps.

The Simplest Trap of All

About 50 years ago, expert Hugo Hartnack recommended the simplest ploy of all. Soak a rag in beer and leave it on the floor overnight near where you've seen roaches. In the morning just step on the drunken bugs.

Commercial Traps

Roach Coaches, Roach Motels, and other such baited boxes attract German cockroaches, fewer of other species. Also they may draw ants, who show up to consume the dead roaches.

None of these traps will work very well if the insects can easily find other food. Good housekeeping makes hungry roaches easier prey.

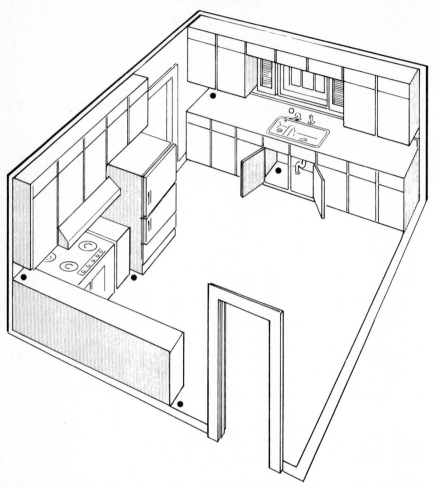

Figure 3.4. Proper placement of roach traps. (*Controlling Household Cock-roaches*. Division of Agricultural Sciences, University of California, Leaflet 21035, 1978.)

Chickens: An Efficient Cockroach Patrol

If you live in a rural or semirural area, consider keeping a few chickens. Apart from eggs and meat, you'll have many fewer vermin around your home. As an old Spanish proverb says, "When arguing with a chicken, the cockroach is always wrong." It's worth a try.

Temperature: Hot, Cold, and Just Right

Temperature can attract the pests or drive them out. Setting your thermostat at a steady 81°F is a welcome mat for roaches, so keep your home on the cool side.

Heat. The British Museum (Natural History) reports that temperature of 130°F has been used successfully in steam-heated British apartment buildings to kill roaches established in the heating system. Radiators were closed and boilers loaded to capacity. According to the Pasadena, California fire department, several hours at such a sweltering temperature pose no fire hazard. However, if you can't close off hot-air grilles of radiators, you run the risk of damaging furniture or musical instruments.

Cold. At the other end of the thermometer, roaches die at 23°F. If you live where winters are cold, and circumstances permit, leave your windows open all day in freezing weather to wipe out roaches as well as silverfish and clothes moths.

"Help! A roach!" most of us feel like yelling when we find one of these vermin. If we can calm down long enough to make sure our homes are not the kind of habitat cockroaches need, we can get rid of them—and the chemicals sold to kill them—for good.

To reach zero cockroach population in your home, you'll need to use several different tactics at one time. However, careful housekeeping must always be at the head of the list. *Good sanitation is still your strongest strategy.*

Future Controls

Scientists at Yale University have recently synthesized a sex lure that sends male cockroaches into a frenzy, drawing them into a trap. The researchers are optimistic about the product, but don't look for it on store shelves for several years. And don't slacken on your home's maintenance. As the United States Navy found, sex lures may not do the job.

Checklist for Cockroach Control

- ✔ Thorough cleaning: bookshelves and closets; behind refrigerator and stove; inside televisions, electric clocks, and radios.
- ✔ Caulk all cracks in interior and exterior walls. Close up openings around pipes.
- ✔ Repair all plumbing leaks.

- Secure splashboard behind sink and countertop.
- Check inside of water meter.
- Pull up ivy or any other broad-leaved ground cover.
- Clean up lumber and other debris near house. Move lumber away from house.
- Check grocery bags, cartons, and soft-drink cartons before bringing them inside.
- Store all food, including pet food, in tightly closed glass, metal, or plastic containers.
- Never leave dirty dishes in the sink for any length of time.

Critters in the Crackers, Pests in the Pantry

A Pennsylvania homemaker decides to serve some fresh corn bread for her family's supper, a treat she hasn't baked in months. Taking an open box from the cupboard, she notices that some of the cornmeal is sticking together in long threads. As she shakes the grains into her measuring cup, out flies a dark-winged moth. Unnerved, the woman throws the box of cornmeal into the garbage can under the sink.

Across the continent, in Oregon, a man, fixing breakfast, pours cracked wheat into boiling water and goes off to set the table. On uncovering the pot, he finds some tiny brown insects, with tooth-like projections behind their heads, floating dead on top of the thickened gruel. He shrugs, skims off the little corpses, and serves his family the hot cereal.

These people have just met two common pests of the home pantry, a meal moth and a sawtoothed grain beetle. Each will probably find more of the insects in the weeks ahead, and most likely in foods other than cereals.

Along with small mammals, pests of stored food destroy more than a billion dollars worth of food in the United States every year. In some undeveloped countries the loss tops 40 percent.

MADISON COUNTY
CANTON PUBLIC LIBRARY SYSTEM
CANTON, MISS. 39046

Besides grains, pantry insects attack legumes, pasta, dried fruits, cheese, and nuts. One whole group likes meats and, by extension, leather and furs as well as fabrics, books, and wood. Another tough little character can gnaw through aluminum and lead.

How They Get In

Before our ancestors learned to store food, these insects lived in nests of seed-eating birds and rodents, in beehives, under tree bark, in wasp nests, and around any cache of seeds. They also ate dead vegetable and animal matter. Our knowledge of food keeping hasn't changed their habits; it's just provided these nuisances with nearly perfect living conditions.

- Chances are you brought your freeloaders into your kitchen with the groceries, maybe in damaged packages of breakfast food or pet kibble. Or you didn't notice that the box of raisins you put in your cart had tiny scrap marks or nearly microscopic holes.
- Other harborages of these pantry creeps are old furniture, drapes, bedding, rugs—any items made with animal fibers.
- Bird and rodent nests as well as nearby dumps and the by-products of food processing plants are also possible sources. These pests have many routes for infiltrating a home from the outside—under doors, through and around torn or loose screens, down fireplace chimneys, and around utility pipes.

Early Signs of Pantry Pests

Weeks, maybe months after the earliest arrivals sneaked in, their numbers have multiplied to the point where you know you have a problem. Your pantry, dark, possibly humid, not too cold, free of predators, and holding a huge stock (by insect standards) of food is an ideal environment for these animals. By the time you notice them, they've probably spread to other packages on your shelf, and the often slow, expensive process of getting rid of them has to begin.

You needn't wait until you see crawlers in the cornflakes to know there's trouble in the kitchen. Early warnings can alert you to a potential or recent invasion. When you find insect-damaged packages on the supermarket shelf, leave them there. And let the store manager know that the warehouse inspection procedures need to be tightened.

If a container of flour or cereal has lumps of food clinging to the sides, you probably have flour beetles. Or the food may be criss-crossed with webbing. By the time this appears, insect young have spun their cocoons, and are resting before emerging as egg-laying adults. Or the food may be dirty with droppings, cast-off larval skins, and partly chewed food particles.

In heavy infestations some insects give off a peculiar odor. Flour beetles have an oppressive, penetrating scent, while crushed mites will remind you of mint.

Four Kinds of Pantry Pests

There are many species of pantry pests, some more common, some more destructive, than others. All can be grouped into four large categories: beetles, weevils, moths, and mites.

1. *Beetles* may be feasting on your flour, dried fruits, cigarettes, prepared cereals, pastry and biscuit mixes, dried soups, herbs and spices, cured meats.
2. *Weevils* go for rice, unmilled wheat, dried peas and beans.
3. *Moths* don't feed on anything, but their larvae devour nuts, dried fruits, cornmeal, other grains.
4. *Mites* like just about all the foods the others eat, plus barley, cheese, wine, sugar, caramel, and rotting potatoes.

Left unchecked, all these anthropods can spread throughout a house, climbing up the walls and thronging around windows. Even confined to the kitchen, they make most homemakers very nervous.

Nature of the Pests

Below are capsule descriptions of the most common wildlife invaders of the home pantry. Figure 4.1 pictures most of them. The section titled "Shut Them Out" details measures for prevention; "Wipe Them Out" tells how to get rid of them once they're in.

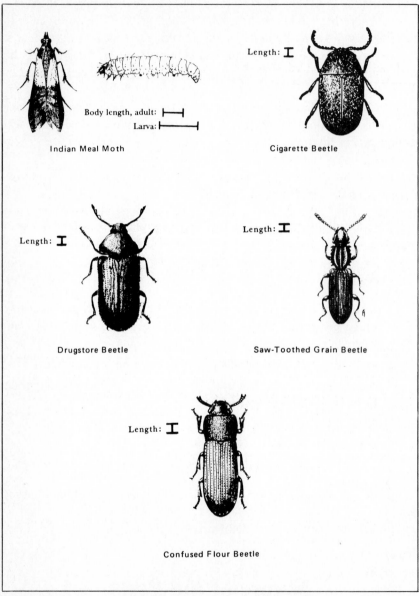

Figure 4.1. Some common pantry pests. (*Common Pantry Pests and Their Control*. Division of Agricultural Sciences, University of California, Leaflet 2711, 1982).

Beetles

The Cadelle

Destructive aboard ship and in grain mills, the cadelle is also a frequent pantry nuisance. It easily gnaws through sacks and cartons, and likes to lay its eggs under loosened carton flaps. Its habit of burrowing into woodwork creates hiding places for other pests.

Appearance: Shiny black, flat; about ⅜-inch long. Head and thorax separated from rest of body by loose, prominent joint.
Susceptible Foods: Cereals, nuts, potatoes, flour, fruits, whole grains.
Signs in Food: Adult insect or larvae.
Prevention: Inspect food packages before purchase, especially for loosened flaps. Rotate food stocks. Buy staples in moderate amounts. Clean cupboards regularly; keep them dry.
Treatment: Discard infested foods; quarantine susceptible ones. Heat or extreme cold on uninfested susceptible items gives more positive control.

Carpet Beetles

Are you surprised to find these fabric destroyers among pantry pests? Carpet beetles frequently attack stored foods. In nature they're primarily scavengers, feeding on animal remains. Keep an eye on any Egyptian mummies you have around the house for these beetles will devour them. Foods these insects contaminate can cause severe digestive upsets and the pests themselves bring on allergies.

Carpet beetles can go without food for a year, a great convenience for a creature that must wait for a stray cadaver to show up before it can get a meal. A dead rodent within a building's walls will attract them.

Appearance: Varies among species. Length runs from about ⅛- to nearly ½-inch. Color can be glossy black to markings of yellow, brown, and white. Larvae look like little bristly carrots.
Susceptible Foods: Dried milk, cayenne pepper, legumes, seeds, corn, wheat, and rice. Adults also feed on pollen.
Signs in Foods: Adult beetles, larvae, cast-off larval skins, fecal pellets. Larvae are active and may be found elsewhere in house. In heavy infestations, adults swarm at windows.
Prevention: Inspect food packages before purchase. Rotate food stocks; buy staples in moderate amounts. Keep whole house, including kitchen

cupboards, clean. Eliminate bird, rodent, and wasp nests near house. Eliminate white flowers in garden as adults seem to prefer these. Check all cut flowers before bringing them inside.

Treatment: Thorough cleaning, including vacuuming. Discard infested foods. Quarantine susceptible foods. Heat or extreme cold gives more positive control.

Cigarette Beetles

This insect is especially fond of tobacco products, but don't think you're safe because you don't smoke. The cigarette beetle attacks a wider variety of food than any other pest of stored products.

Appearance: About $\frac{1}{10}$-inch long; reddish brown; covered with short, fine hair. Low-hanging head gives it humped appearance.

Susceptible Foods and Other Items: Herbs, spices, grains; dried fruits, meats, fish, vegetables; nuts, coffee beans. Leather, insecticides containing pyrethrum (this is some tough cookie!), furniture stuffing, bookbinder's paste. Paprika and dog food are most often infested in the home.

Signs in Food: Insects themselves and larvae. Tiny holes in containers. Webbing in tight places.

Prevention: Inspect food packages before purchase. Keep cupboards clean and dry. Rotate food stocks and buy staples in moderate amounts. Refrigerate susceptible foods in warm weather.

Treatment: Discard infested food and other items. Quarantine susceptible foods. Use heat or extreme cold on susceptible items for more positive control.

Drugstore Beetles

One of the toughest pantry pests, the drugstore beetle eats anything except cast iron. It has the eerie ability to thrive on poisons like aconite, belladonna, and strychnine.

Appearance: About $\frac{1}{10}$-inch long. Light brown and covered with silky down. Low-hanging head.

Susceptible Foods and Other Items: Any household foods, spices, bird seed, pet foods, pastry mixes. Drugs, books, tin, aluminum foil.

Signs in Foods: Adult beetles, larvae. Cocoons covered with bits of food. Webbing in tight places. Tiny holes in containers.

Prevention: Inspect food packages before purchase. Rotate food stocks. Buy staples in moderate amounts. Clean cupboards regularly and keep them dry. In warm weather refrigerate susceptible foods.

Treatment: Discard infested foods. Quarantine susceptible ones. Heat or extreme cold gives more positive control.

Flour Beetles

Along with the sawtoothed grain beetle and the Indian meal moth, these are the most common pests in markets and the home pantry. Clever humans have found some use for them. Because they can't climb glass, they make docile laboratory animals. And a thousand of them, along with some tiny wasps, were the first earthlings rocketed into space.

Appearance: About $\frac{1}{7}$ of an inch long; flat oval shape, reddish-brown.

Susceptible Foods: Grains, legumes, shelled nuts, dried fruits, chocolate, spices, snuff, cayenne pepper. They can also ruin museum and herbarium specimens.

Signs in Food: Clumps of food and eggs clinging to sides of container. Tiny holes in packages. Webbing in tight places. Adult insects and their droppings. Give off foul odor as adults.

Prevention: Inspect food packages before purchase. Rotate food stocks and buy staples in moderate amounts. Vacuum pantry shelves regularly. Keep pantry dry.

Treatment: Discard infested foods. Quarantine susceptible ones. Heat or extreme cold give more positive control. If only eggs are present in a large quantity of flour that you want to save, sift the flour through bolting cloth. This is a fine-meshed silk or nylon used by grain processors for the final milling. Also used by artists and photographers, the cloth can be bought in art and photography supply stores.

Sawtoothed Grain Beetles

Sometimes mistakenly called the sawtoothed weevil, this tiny beetle resembles a miniature double-edged saw. It's a lively little cuss whose flat shape lets it penetrate sealed packages. It can't damage whole grain, but once this is milled or otherwise broken, the insect mounts its assault.

Appearance: Bright brown; about $\frac{1}{10}$-inch long. Has six little "teeth" on each side of thorax.

Susceptible Foods: Cereals and other grains, spices, tobacco, meats, candies, dried fruits. Is especially fond of raisins.

Signs in Foods: The adult beetle, webbing in tight places, tiny holes in packages.

Prevention: Inspect food packages before purchase. Rotate food stocks and buy staples in moderate amounts. Vacuum pantry shelves regularly. Keep cupboards dry. Store susceptible foods in refrigerator.

Treatment: Discard infested foods. Quarantine susceptible ones. Clean pantry thoroughly. Heat or extreme cold gives positive control for susceptible foods.

Spider Beetles

Its long legs give this beetle its name. The spider beetle infests a vast range of stored products, including foods, only some of which are listed below. Because this insect can survive very low temperatures, refrigeration does little to control it.

Appearance: Depending on species, length can be from less than $\frac{1}{16}$- to nearly $\frac{1}{4}$-inch. Long legs. Color ranges from red to various shades of brown, black, or yellow.

Susceptible Foods and Other Items (partial list):   Almonds, cocoa, corn, dried soups and fruits, rye, spices, fish food, raisins; casein, furs, feathers, silk, linen, packaging material, wood, lead, drugs, excrement.

Signs in Food: Adult insects, larvae, food lumped around cocoons.

Prevention: Inspect food packages before purchase. Clean cupboards regularly. Rotate food stocks and buy staples in moderate amounts. Keep cupboards dry.

Treatment: Discard infested foods. Quarantine susceptible ones. Heating uninfested susceptible foods gives more positive control.

Flour and Grain Moths

Major pests in grain mills, these moths can do much damage in the home. The adults are harmless but their larvae spin cocoons in our food and otherwise contaminate it. They're extremely destructive.

On first spotting one of these, you may wonder what a clothes moth is doing among the groceries. Don't be fooled. A clothes moth is a drab buff color; a meal moth is handsome, with either coppery or dark gray bands striping pale gray wings.

Appearance: Adult wingspread about $\frac{3}{4}$- to $\frac{1}{2}$-inch. Copper-colored or gray stripes on wings. Larvae about $\frac{1}{2}$-inch long; color varies from dirty white to pinkish or greenish hues.

Susceptible Foods: Bran, biscuits, dog food, nuts, seeds, chocolate, crackers, powdered milk, candy, red peppers, dehydrated vegetables, dried fruit. If you find "white worms" in dried fruit, they're probably grain moth larvae. The larvae can also be found in other parts of the house.

Signs in Food: Widespread webbing. One species, which infests only whole grains, imparts sickening odor and taste.

Prevention: Inspect food packages before purchase. Buy moderate amounts of staples and rotate food stocks. Refrigerate whole grains. Vacuum cupboard shelves and walls regularly. Keep pantry dry.

Treatment: Discard infested foods. Quarantine susceptible ones. Purchase replacements in small amounts. Heat or extreme cold will give more positive control of susceptible foods.

Weevils

Granary and Rice Weevils

These pests may be found anywhere in the house. Their long, hard snouts enable them to bore through hard substances. In cooler climates the granary weevil is more apt to invade your pantry; in warmer regions the invader will be the rice weevil. Rice weevils can fly, granary weevils cannot.

Appearance: brownish or blackish color, about $\frac{1}{8}$-inch long. Long, blunt-ended snout. Half-moon shaped, legless larvae.

Susceptible Foods: Whole grains, pasta, beans, nuts, cereal products, grapes, apples, pears.

Signs in Food: Adults and larvae. Holes bored in foods.

Prevention: Inspect food packages before purchase. Rotate food stocks and buy staples in moderate amounts. Vacuum pantry shelves regularly.

Treatment: Discard infested foods. Quarantine susceptible ones. Heat or extreme cold gives more positive control.

Bean Weevils

Not a true weevil since it lacks the long snout, this insect has been a major pest throughout recorded history. It has been found in beans in ancient Inca graves and is mentioned in the Talmud.

Appearance: Velvety gray or brown. Long black and white markings. About ⅛-inch long. Chunky shape with rear wider than head.
Susceptible Foods: Beans, peas, lentils.
Signs in Food: The adult beetle. Holes in legumes.
Prevention: Inspect legume packages before purchase. Keep cupboards cool. Buy legumes in moderate amounts. Rotate stocks.
Treatment: Discard infested items. Quarantine susceptible ones. Heat or extreme cold gives more positive control of susceptible foods.

Mites

Like their relatives, ticks and scorpions, mites spell trouble for humans. They carry tapeworms and cause kidney and skin problems. Grocer's itch, a severe dermatitis, is caused by mites.

Mites spread very quickly from bird and rodent nests. You can easily pick them up and, unawares, carry them home on your shoes and clothing.

Appearance: Almost microscopic in size, translucent. Sparse body hairs. Move slowly away from light.
Susceptible Foods: Dried fruits, flour, meat, cheese, grains, mold on foods, dried bananas, dried milk, rotting potatoes, caramel, fermenting substances, nuts, mushrooms.
Signs in Food: Buff-colored, pinkish, or grayish dust around food. Unpleasant or minty odor. The animal itself.
Prevention: Discourage birds building nests near your home. Rotate foodstuffs and buy staples in moderate amounts. Discard molding foods. Clean cupboards, especially horizontal surfaces, regularly. Keep cupboards dry.
Treatment: Discard infested foods. Heat uninfested susceptible foods. Clean cupboards thoroughly.

The Cheese Skipper

This oddball acrobat is the maggot of a fly. Like the young of any filth fly, it can cause serious digestive illness. Fastening its mouth hook on its abdomen, and then suddenly releasing it, the maggot can leap 10 inches across a surface.

Appearance: Adult fly is small, black or bluish-black with bronze tints on thorax and face.

Susceptible Foods: Cheese, pork, beef, smoked fish, brine-cured fish, animal carcasses.

Signs in Food: Presence of maggots and flies.

Prevention: Careful sanitation, tight-fitting screens of 30-mesh wire or plastic. Cheese can be protected by wrapping in cheesecloth and dipping in paraffin. Certain types of bags protect ham. Cover and refrigerate all meat and cheese.

Treatment: Discard infested foods.

Shut Them Out

Prevention: The Easiest Control

Halting pantry pests before they invade, settle in, and raise their families in our foods is a lot easier than trying to evict them. Here are ten simple routines that will keep insects and mites out of your cupboards.

Begin Outside and Follow Through Inside

1. Eliminate any bird, wasp, or rodent nests near your house. These often harbor stored-food pests. If a pair of birds starts carrying twigs and lint to a vine growing on your wall or a tree overhanging the roof, chase them off. Watching a family of nestlings develop is fascinating, but it's no fun to cope with the pests that can migrate from their nest to your kitchen.

2. Before putting any packaged foods into your supermarket cart, inspect them carefully. If package flaps are loose or the cardboard is punctured, no matter how tiny the holes—especially if the holes are tiny!—don't take the item. Be especially fussy about all cereals, pasta, pastry, cake, biscuit mixes, dried fruits, legumes, powdered milk, bird seed, and kibbled pet food.

    If you find a damaged package when you get the groceries home, take it back immediately, before any possible infestation has a chance to spread.

3. Be careful when buying used furniture, bedding, or drapes. The animal or vegetable fibers these often contain may be sheltering pantry pests.

    Indoors

4. Keep your kitchen cool and dry. If possible, ventilate your cupboards. Many stored-food pests need a humid environment and temperatures between 75 and 85°F.

Packets of silica gel on pantry shelves help hold down humidity. Silica gel can be bought at builders' supply stores. Set the granules in a jar covered with a punctured lid. When they've turned pink, they've absorbed all the moisture they can. Dry them out in the oven until they've turned grayish-white again.

5. Make sure your kitchen walls, ceilings, and floor are free of cracks and holes where insects can hide and lay their eggs.

6. Kitchen drawers and cupboards should be kept clean. Canisters should be washed thoroughly inside and out before refilling. Vacuum the insides of your cupboards at regular intervals to pick up crumbs. Don't overlook the undersides of shelves, where pest young often hang their cocoons. Occasional scrubbings with soap or detergent and water also help.

7. Rotate your food stocks, using up older ones first. Don't buy more grain foods, especially in the summer, than you'll consume in a short time. If you don't bake or eat hot breakfasts in the summer, use up cereals and baking supplies before hot weather sets in.

Note: It's probably safe to keep sound, unopened boxes of cereals on the shelf in warm weather because in most cases cartons have been treated with a repellent approved by the Food and Drug Administration.

8. Don't mix new foods with old foods or you could be spreading an infestation from one to the other.

9. Check your shelves from time to time for opened, forgotten packages. These are often the major harborage for these pests. Cereals that have only one insect in a box may be swarming with them a month later. In a well-run food warehouse, inspection of stock is ongoing and at the first sign of wildlife, the food is destroyed or treated.

Since the reproductive cycle of some of these insects can be less than a week when temperatures are high, check your staples at least once a week in summer, once a month in winter. While looking over each item, shake or stir it vigorously to disrupt any reproductive process going on.

10. Store all nonrefrigerated foods in tightly closed containers. Be especially careful about pet food, a favorite with many pantry pests. A tightly lidded garbage can locks them out of kibble; for smaller amounts, a tin saltine box works fine. Spices and herbs should be in screwtop jars or reclosable tins.

Herbal Repellents

In times past, careful homemakers routinely dropped a bay leaf or two into their stocks of grains to repel insects. Actually, this may keep out a stray bug or two but don't rely on it to suppress insects that come into your home with the groceries. Some of these pests, like the eat-everything cigarette beetle, may like bay leaves—or cinnamon, cloves, mint, and any other reputed repellent. The homemakers who found herbals reliable probably kept their pantries immaculate.

However, if bay leaves in the cereal make you feel more secure, put them into a cloth bag. Otherwise the brittle leaves will crumble into the food. And who wants to eat bay-flavored oatmeal?

A Grain Mill That Uses No Chemical Pesticides

In the Texas panhandle, not far from Amarillo, Arrowhead Mills processes tons of grains and legumes, much of them stored in outside bins, all year round without using pesticides. How do they do it? According to the company's president, Boyd Foster, with scrupulous cleanliness, cold, and mechanical jolting.

Before each new batch of grain is poured into the machinery, all milling equipment is thoroughly cleaned and vacuumed. Storage warehouses are kept at a steady 40°F. The company also has an ingenious solution to the problem of stored-food pests in its outdoor bins, where unprocessed foods are kept for months. Fans at the bottom of the bins run all through the frigid panhandle winter, pulling icy air through the mass of grains and legumes. So deep is the chilling that the food remains protected all during the rest of the year.

Harvested grains and beans often carry insect eggs that don't hatch until the food is in milling or storage, conceivably even until it reaches your kitchen. To destroy such eggs, Arrowhead Mills have raised the stirring and shaking of stored foods to a fine art. With a device called an "entoleter," they spin the stocks at a dizzying speed, ending with a sharp jolt. Any insect eggs are completely crushed.

Wipe Them Out

Getting Rid of Pantry Pests

1. **Persistence and watchfulness.** As soon as you notice any sign of these insects, look over all the foods in your cabinets, including any unopened cardboard boxes or cellophane or plastic bags. If you find even one insect, its webbing or larvae; or tiny punctures or scrapings on the outside of a container, put the food in a plastic bag and close it securely with a knot or rubber band; put the bag in a garbage container outside—preferably in the hot sun so the heat will kill any organisms.

2. **Thorough cleaning.** Empty your pantry completely and vacuum and/or wash all surfaces thoroughly. Use your crevice tool to suck up bits of food that have fallen behind shelves and into cracks.

3. **Quarantine.** Place any suspect foods—those near the infestation—in plastic bags for a few days so you can watch them for signs of insect activity. If life seems to be developing in any of them, throw them out as indicated above. Quarantine food in unopened packages by putting the whole package into a plastic bag. Any insects emerging from the container will be visible but unable to penetrate the plastic.

4. **Cold.** If your refrigerator is large enough, store the rest of your opened food stocks in it for a few weeks. Most species die after prolonged exposure to 40°F. (An exception is the spider beetle, which can withstand great cold.)

    If you have a deep freeze with a steady temperature of 0 to 5°F, you have an all-purpose pesticide. Two to three days in deep cold kill most pest species, and three weeks will destroy all their life stages.

5. **Heat.** Lacking a freezer, use heat. Spread your staples in thin layers on flat pans with raised edges. Heat them for 2 hours in your oven at about 125°F. Don't let the heat rise much over that or you'll destroy some nutrients. Just turning the pilot light up a bit may do the trick. Use an oven thermometer to check the temperature from time to time. If it rises over 130°F, prop the oven door open a few inches to bring it down.

6. **Boiling water.** If you find wildlife in raisins, prunes, apricots, or

other dried fruits and you want to salvage these expensive foods, drop them in boiling water for 1 minute. Rinse them off and dry them before putting them into a tightly closed container.

7. **Carbon dioxide.**   Carbon dioxide can be used to kill any organism in your food. Dry ice, which is solidified CO_2, is usually available at ice-making plants. Some ice cream stores may also sell it.

(*Caution:*   You can get a bad freezer burn from dry ice. *Do not handle it with bare hands.* Wear thick gloves or use tongs.)

Put a layer of the food you're treating in a clear plastic container. If you use glass, don't let the dry ice touch the sides of the jar as the extreme cold can crack it. Place a cube of dry ice on the food and fill the container with more of the food. A ½-inch cube is enough for a pint jar, a 1-inch cube clears out a quart jar, and a 2-inch piece takes care of a 2-quart container.

If the lids are screw tops, turn them until they begin to tighten; then turn them back to loosen them. *Lids must be loose to allow gas to escape and prevent an explosion.* Tightly enclosed, dry ice can build up a pressure of 200 to 250 pounds per square inch, enough to burst a glass jar.

If your containers are plastic with snap-on lids, "burp" them every once in a while by lifting a lid corner briefly to hold down the pressure inside.

After several hours, when the dry ice has evaporated, seal the container completely. Treating foods with carbon dioxide does not change their quality. Cereals and grains do not lump.

Pantry pests can drive a homemaker frantic. When you think you're finally rid of them, they turn up in food you're positive was free of them just a short time before. Food is expensive and throwing it out painful. You may be desperate enough to consider spraying your pantry with a chemical. Don't do it. No insecticide is safe for such a use. And remember, some of these animals thrive on poison.

Your best protection against these mini-pirates is prevention—watchful food buying; keeping your home clean, cool, and dry; buying food in moderate amounts; and keeping a close eye on your pantry and its contents.

If some insects manage to slip through your defenses, the safest way to get rid of them is to:

• Throw out all infested items.

• Quarantine all susceptible foods not showing signs of insect activity.

- Heat or freeze all susceptible foods.
- Buy new stocks in the smallest amounts possible for a few months, and use them up quickly. Keep them completely separate from foods already in your kitchen.
- Hang in there! It may take time but eventually you'll be rid of your pantry pests.

CHAPTER 5

Rats and Mice: There's No Pied Piper

On a cold day in December 1818, Pastor Joseph Mohr sat down at the church organ in Obendorf, Austria, to practice for the Christmas Eve mass. He pressed the keys and pumped the pedals, but the sturdy old instrument gave out more rheumatic wheezes and sighs than musical tones. Mice had chewed the bellows full of holes! Having no time to repair the organ and determined not to have the mass without music, he asked his friend Franz Gruber to compose a simple melody for guitar while he wrote the lyrics. They worked through the night of December 23, and on Christmas Eve, the two friends gave the first public performance of "Silent Night, Holy Night."

Except for a distant relative named Mickey who's made it big in Hollywood, that was probably the only worthwhile thing a domestic rodent has ever done for humanity.

Mice and their larger kin, rats, have plagued us for thousands of years. They have invaded our dwellings, food stores, museums, offices, transport vehicles—even our books. So intimately are these animals bound up in our lives, they're called "domestic" rodents. If you object to mice and rats sharing billing with your dog or cat, you could use the term "commensal," meaning they share your table—a notion you may like even less.

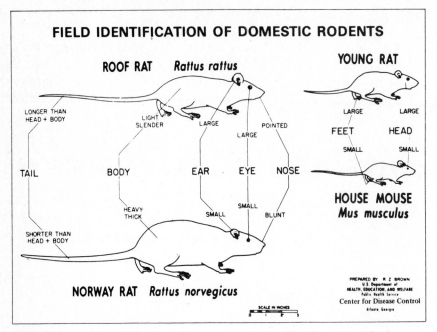

FIELD IDENTIFICATION OF DOMESTIC RODENTS

ROOF RAT *Rattus rattus*

YOUNG RAT

LONGER THAN
HEAD + BODY

LIGHT
SLENDER   LARGE   POINTED

LARGE

LARGE   LARGE

FEET   HEAD

SMALL   SMALL

TAIL   BODY   EAR   EYE   NOSE

HEAVY
THICK   SMALL   SMALL   BLUNT

SHORTER THAN
HEAD + BODY

HOUSE MOUSE
Mus musculus

NORWAY RAT *Rattus norvegicus*

PREPARED BY   R Z BROWN
U.S. Department of
HEALTH, EDUCATION, AND WELFARE
Public Health Service
Center for Disease Control
Atlanta, Georgia

SCALE IN INCHES

Figure 5.1. Field identification of domestic rodents. The Norway rat is the most common in temperate zones; the roof rat in the tropics and subtropics. The house mouse thrives everywhere. (*Control of Domestic Rats and Mice*. Atlanta: Centers for Disease Control, Homestudy Course 3013-G, Manual 11, 1982.)

Among the most successful mammals, except for humans, mice are the most widely distributed of this major group. Both rats and mice probably originated in Asia but they've been carried in ships, caravans, trains, trucks, even airplanes, to the four corners of the globe.

In the United States the roof rat is the one most likely pattering over wires and fences along the West and Gulf coasts, and in the southeastern states as far north as North Carolina. The Norway rat (also known as the sewer, wharf, or brown rat) and the house mouse are the most common rodent pests throughout the rest of the country. The larger, heavier Norway rat is the one generally found in human structures; the lighter, more agile roof rat can prosper independently of human habitation. Figure 5.1 illustrates the differences in appearance of the three species.

Albino rats, used in many laboratories and often kept as pets, are Norway rats. When they escape or are intentionally released, they quickly establish themselves as community pests, and are as destructive and dangerous as their darker cousins.

Dangers from Rodents

Throughout history domestic rodents have tormented us. Bubonic plague, the dreaded Black Death, is carried by rat fleas. It has killed millions of people since at least the sixth century A.D. The Pied Piper was said to have charmed away Hamelin's rats in 1284, perhaps the beginning of the plague epidemic that nearly depopulated Europe in the following century. Despite effective antibiotics, thousands still die of it every year.

Other diseases rats and mice can give us are typhus, dysentery, food poisoning, trichinosis, and ratbite fever (despite its name, also transmitted by mice). Dale Bottrell, author of *Integrated Pest Management,* says that an estimated 60,000 Americans, mostly infants in their cribs and bedridden elderly, are bitten by rats every year. More vicious than mice, rats will attack if cornered, and can chew off fingers and toes. Once they've bitten a person, they will probably do so again.

Economic Damage Caused by Rats and Mice

The economic harm these animals do is almost beyond counting. Rats must gnaw constantly to keep their incisors, which grow about an inch a year, from piercing their skulls. They gnaw into upholstery and other fabrics, chew books and paper to get nesting materials. Filing their teeth on electrical wires and matches, they set homes on fire. Exerting a massive 24,000 pounds per square inch with their incisors, they can gouge holes in gas pipes, asphyxiating whole families.

And mice are nearly as destructive. Besides ruining crops and food stocks, they nibble family records, valuable paintings and manuscripts. Because it has no bladder, a mouse urinates constantly, contaminating everything it touches. It can make itself at home in an electrical appliance, causing dangerous short circuits.

Nature of Rodents

Where and How Rats Live

Both the Norway rat and the roof rat find human habitation very comfortable. With its tendency to burrow, the Norway rat is ideally suited to a rural environment. The roof rat, an agile climber, is more at home in the city with its wires and tall buildings.

The Norway Rat:

* Burrows under buildings and piles of lumber to travel to and from its nest. Burrows are up to 15 inches deep, 2½ to 4 inches wide, and may be as long as a city block. Bolt holes, set under grass or boards, provide rodents with a quick escape route.
* Nests in damp places—stream banks, garbage dumps, marshes.
* Eats both vegetables and meat and is smart enough to prefer fresh food to spoiled.
* Is nocturnal.
* In its 9- to 12-month life span can produce four to seven litters of eight to twelve young each.

The Roof Rat:

* Nests in woodpiles, pyracantha bushes, thick growths of ivy, palm trees and other evergreens, all found in the suburbs.
* Eats insects, berries, grains, nuts, fruits, vegetables, manure. Needs less water than Norway rat.
* Is nocturnal.
* In its 9- to 12-month life span, produces up to six litters, with six to eight young each.

Both Norway and roof rats are strong swimmers, able to paddle around for hours, and are often found in urban sewers. Even in affluent neighborhoods they can swim through floor drains and toilet water seals. In many large cities rats use older sewage systems for highways. When storm and sanitary sewers are combined, the sewer rat problem is apt to be worse, especially if garbage disposers are in general use and food scraps mingle with waste water.

Rats rove anywhere from 100 to 150 feet from their nests; mice about 10 to 30 feet from theirs. Between nests and food sources they establish harborages, resting places where they can hide and eat. Neither species sees well so they tend to run along a wall, where their long guard hairs make contact and guide them.

Where and How Mice Live

Because the mouse originated in the dry grasslands of Asia, it's well adapted to life without a steady water supply, and is able to survive long periods in closed boxes, trunks, and barrels, where it has plenty of time to wreak havoc. Like rats, mice are very good swimmers.

The House Mouse

- Can live comfortably in a sack of grain or other dry food without visible moisture.
- Burrows, but its tiny trails are impossible to find.
- May produce seven or eight litters in its lifetime, with five or six young each.

Mice are even more adaptable than rats to life with human beings. They've been found flourishing in the upper stories of high-rise apartment buildings even before the structures were finished, probably living on workers' lunch scraps.

Exploding a Myth: Rats Abound in the Suburbs

On first moving to the suburbs, former city people are often delighted with their new neighbors—deer, rabbits, cedar waxwings, and racoons. Their delight may quickly fade, however, when they come upon snakes, skunks, rats, and mice also living in the vicinity.

Many city dwellers live in apartments all their lives without ever meeting up with a rat or mouse. Although blighted urban neighborhoods carry large rodent populations, well-kept buildings are usually free of such wildlife.

Suburban Infestation Patterns

Rodent infestation of suburban housing follows a marked pattern. Many of these developments are built on former orchards or, in the sun-belt, on citrus groves, where rodents may be plentiful. For a few years the new homes are rodent-free, but then pools are put in; bird feeders erected; shrubs, trees, and vines thrive; and the stacks of items stored in the attached garages, basements, or attics grow. Shake roofs develop holes; vent screens deteriorate; the house settles, opening small cracks under the eaves, in the foundation, and around doors and windows.

Within 10 to 12 years the roof rat population explodes, and as many as 20 to 25 percent of the homes are infested. Twenty to forty years later, with trees and shrubs mature and some of the homes poorly kept, the whole area is open to vermin infestation.

The most vulnerable homes are older, neglected, and with many points of entry and hiding places for small animals. Poor disposal of food waste aggravates the problem, as do nut trees, outbuildings, wood piles, and thick vegetation.

Signs of Rodents

"Out of sight, out of mind" describes most people's attitudes toward domestic rodents. If we don't see them, we think they aren't there. But there are clues to their presence that should make you spring into action. Either one or several of the following signs can mean that you have some four-footed boarders.

Sounds

At night you hear pattering of paws, climbing sounds in the walls, squeaks, churring sounds. Mice give a little whistle, rather like a canary. To be positive that these are rodent noises, after you've been out for an hour or two in the evening, enter your house very quietly and stand still for a little while. If rodents are in the walls, their sounds will gradually resume.

Droppings

By the time you hear rodent noises, you may have noticed droppings in your kitchen cupboard or on the countertop. They'll be heaviest along rat or mouse runways and near your food supplies. Variations in size can indicate an established colony, with older and younger animals.

Some authorities recommend flashing black light in the area to detect rodent urine, which fluoresces under ultraviolet light. However, so do other substances so this is not an infallible clue. In long-term infestations mice leave a "urine pillar," a pile of urine, grease, and dirt.

Gnawings

Freshly gnawed wood, lighter in color than the wood around it, and with small chips of the same color scattered about, are a strong rodent sign. Since these animals will smooth over a hole through which they have to pass, scratchy, sharp gnawings indicate recent activity.

Grease Marks

Greasy stains from rodents' fur along runways and where they swing past the angles of joists and ceiling rafters are another telltale clue. Such marks can also rim gnawed holes, follow along pipes and beams, and soil the edges of stairs. Mouse smudges are much smaller and harder to see. The Norway rat runs along the floor so smudges higher up indicate roof rats.

Odor

In large numbers, rats and mice give off a peculiar pungent smell. If your house has a persistent odor you can't identify, you'd be wise to look for rodents.

An Excited Pet

When your cat or dog repeatedly sniffs and paws at the floor in front of one wall spot, the kitchen cabinet, or among stored items in the garage or basement, you should, as the saying goes, "smell a rat." This is especially true if the invasion is recent.

Sighting a Rat or Mouse

You know you have a problem if you see a live rodent. Rats are secretive and nocturnal, so if you spot one in the daytime, there's a big population of them on the premises. The burrows and nests are crowded and that individual has been forced out. For every rat you see at night, there are ten or more in the general area. Mice are active in the daytime so one in your kitchen in the morning may or may not mean that you have a colony in the house.

Dead Rodent

If you find a dead rat or mouse in or near your home (assuming you don't have a cat that likes to bring you "presents"), *notify public health authorities at once*. Ask if a poisoning program is under way. If the answer is no, the animal may be diseased. *Do not handle it with bare hands*. Use tongs or rubber gloves, or two sticks to put it into a plastic bag. Tie a knot at the top of the bag and throw it into the garbage or bury it. For further details, see the section called "Carcass Disposal."

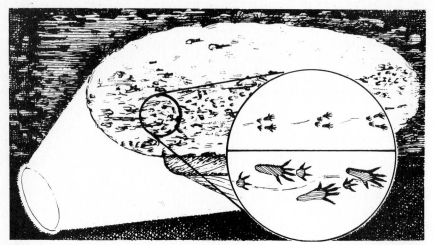

Figure 5.2. Tracks of mice and rats. (*Control of Domestic Rats and Mice.* Atlanta: Centers for Disease Control, Homestudy Course 3013-G, Manual 11, 1982.)

Tracks

Outdoors, rat runways are clean, packed earth paths 2 to 3 inches wide, with tracks of clawmarks in the dust. They're readily seen along walls, under boards, and behind stored objects and litter. You may also notice paw marks in the mud around a puddle. It's easier to see tracks at night with a flashlight held sideways so that the indentations cast a shadow. See Figure 5.2.

If you suspect there's a runway outdoors but the marks are not clear, set a pipe or long narrow box, with its ends cut off, along the path and dust the inside of it with flour or talc. The powder will be protected from the elements and any rodent running across it will announce itself.

To detect rodent paths indoors, sprinkle a nontoxic powder like flour or talc along where you think the animals are traveling. If you've spotted a runway, you'll soon see giveaway tracks.

Feeding Stations

Rodents often have a favorite place to eat away from the nest. Accumulations of bits of gnawed bones, seed hulls, food wrappings, and the like will mark such a spot.

Immediate Control

Your first reaction to these rodent signs may be to set out an array of traps or poison baits. Used skillfully and in sufficient quantities, these can quickly reduce the number of rats and mice on your property. However, without rigorous sanitation in a short time the survivors' birthrate will soar, reduced competition for food will let more young reach maturity, and you'll soon have as many vermin as before.

Starve Them Out

Sanitation

Tightening sanitation, then, is your first priority. Sanitation includes cleaning out all possible rodent harborages as well as proper handling of stored foods and food waste. By removing harborages and food you intensify survival pressures on the animals. Competition for food becomes fierce, dominant rodents kill weaker ones outright or throw them out of the community so that they become easy meals for predators.

Indoor Sanitation

Give them no rest. To reduce available nesting places inside, clean out the debris in your attic, basement, and closets. Piles of old clothes forgotten in a corner, old papers and magazines, boxes kept for no particular reason should all be set out with the weekly trash. Areas under porches, stairwells, and outside steps frequently collect such rubbish so be sure to clean these out too.

Store all foods in tightly closed metal, glass, or plastic containers. Cupboards should be kept clean and floors swept regularly. Wrap leftovers carefully and store them in the refrigerator. A cake serving storage platter with a lock-on lid protects leftover pastries.

Wipe up all spills promptly, especially milk. When your baby falls asleep with a bottle, take the bottle away. Rats like milk and babies are frequent victims of rat bites.

Outdoor Sanitation

Rats hang out in piles of lumber, dense vegetation, stone rubble, dead palm fronds, and ill-kept storage sheds. Stack building lumber and fireplace logs 18 inches off the ground so that even if rodents are living among

ONE-CAN STAND MADE OF REINFORCING
STEEL WELDED TOGETHER

Figure 5.3. One-can stand made of reinforced steel welded together. (*Rodent-borne Disease Control through
Rodent Stoppage.* Atlanta: Centers for
Disease Control, 1976.)

the logs and planks, you can see their droppings and take control measures.

Cut any vines way back and keep shrubs well trimmed. Pull up weeds
and dead plants and put them in plastic bags for the sanitation truck. If
you have palm trees, have the dead fronds cut down; get rid of Algerian
ivy, a perfect cover for rodents.

Put all your kitchen scraps in a plastic bag kept closed in your garbage
can in the house. At the end of the day, close the bag securely, either by
knotting it or with a rubber band, and dispose of it in a tightly closed
container outside.

Refuse containers should be of 20- to 30-gallon capacity, with handles
for easy lifting and lids that fit securely. They should be made of heavy-
duty plastic or galvanized metal, with recessed bottoms so that no filth
accumulates underneath. A 55-gallon drum isn't suitable as it's too heavy
to lift when filled and its lid usually can't be tightly closed.

Store containers on metal or wooden racks or holders 18 inches above
the ground. Racks can be made of steel pipe or bars or weather-resistant
wood. Racks and holders like those pictured in Figures 5.3 through 5.5
minimize the chance of a roving dog's knocking cans over, loosening lids,
and making the contents available to the local rodent community.

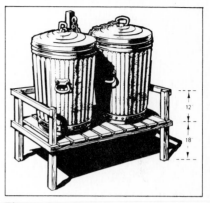

Figure 5.4. This chained refuse container is dog-proof. (*Control of Domestic Rats and Mice*. Atlanta: Centers for Disease Control, Homestudy Course 3013-G, Manual 11, 1982.)

Figure 5.5. Proper rack storage of refuse cans. (*Control of Domestic Rats and Mice*. Atlanta: Centers for Disease Control, Homestudy Course 3013-G, Manual 11, 1982.)

Keeping trash containers on concrete slabs is an effective control measure only if you keep the slab clean, avoid spillage, and always keep lids on tight.

Because rats and mice feed on fruits and nuts that fall from yard trees, keep all dropped fruit picked up. If you have pets—hamsters, guinea pigs, pigeons, dogs, cats—store their food in closed metal or heavy plastic containers.

Snails also are good eating for rats. See Chapter 12 for ways to control snails in your garden, making it less attractive to rats.

If you feed your dog outside, dispose of any leftovers as soon as the animal's finished eating. Make sure no pet food is lost under the dog house, where a rodent can easily reach it by burrowing. Does your dog have a weekly bone or two for keeping its teeth healthy? Rodents also like to gnaw on bones; so, when the dog has finished, throw the bones out.

Birdhouses are a frequent lure for rodents, who eat both the birds' food and their young. Don't put out more than the birds will eat in one day and run a band of Tanglefoot around the birdhouse pole to stop the predaceous rats. And if you enjoy scattering crumbs on the lawn for birds in winter, remember that you may also be helping rodents to survive.

Apartments

If you live in an apartment building where one or two tenants are slobs, your own control measures may not be enough. In such a case, alert the management. It's their responsibility to see that the premises are rodent-free. Fliers and tenant meetings may be all the pressure needed to get things in shape.

Building managers should also see that incinerators work properly at all times. When they only partly consume the refuse, the half-charred food scraps can support many rats and mice, along with the inevitable cockroaches and pantry pests.

Wipe Them Out

With sanitation in and around your home leaving no rodent food or hiding places, it's time to eliminate those still infesting your property. Trapping is your safest, surest method—safest for you and surest to kill the rats or mice. You use no dangerous rodenticides, you quickly see how successful you've been, and you won't have any dead rodents in places where you can't dispose of the bodies—in other words, no dead-rodent odors.

The Snap Trap Is Best

One of the most effective devices ever invented for killing rodents is the well-known snap trap, also called the guillotine, spring, and break-back trap. A large California pest control company relies almost exclusively on such traps, monitored by computers, and rigorous sanitation to keep commercial baking plants and other food processing facilities rodent-free.

Choose the Right One

Be sure to get the right size for the rodent you're trying to catch. The spring of a rat trap can completely miss a mouse taking the bait, while a mouse trap won't do much to a rat. Once you've found the right size, choose the model with the strongest spring and most sensitive trigger.

How Many to Buy

Buy enough to mount a quick, decisive campaign. For the average-size home, twelve or so will do the job. For a farm, you may need anywhere from fifty to a hundred. A reasonable rule of thumb is to use more traps than you think you have rodents. Fewer traps used for a longer time will simply make the survivors trap-shy.

Effective Bait

Your choice of bait is critical. Despite the folklore, cheese doesn't work too well, and exterminators usually scorn it as a lure, For Norway rats, use a piece of bacon or a slice of hot dog; for roof rats, nutmeats, raisins, or a prune. Mice are lured by either a gumdrop or a piece of bacon. Peanut butter smeared on the trigger will attract all species. All the baits can be made more enticing if they're sprinkled with a bit of oatmeal or cornmeal. Tie the bit of food firmly to the trigger with a piece of light string or fine wire, or a smart rodent (and some are very smart) may get the free meal without being caught.

You may need to try several baits before you find one that works. Baits are most effective when there's little other food around, so keep up your good sanitation routine. To prevent rodents from becoming trap-shy, don't set the baited traps until the bit of food has been taken at least once.

Trap Placement

Before placing the traps, sprinkle runways and harborages with flea powder to kill the parasites that will leave the dead rodents and move out to nearby humans or their pets.

1. Set traps close to the walls where your tracking powder has shown you that the rodents run, and behind objects and in dark corners.

2. Position traps so that in the natural course of foraging, the rodent must pass over the trigger. If the trap is set along a partition, it should extend out from the wall and have its trigger nearly touching the wall. If placed parallel to the wall, use a pair of traps with triggers placed so as to intercept animals coming from either direction. (See Figure 5.6A.)

3. To guide the rodents into the traps, set a box near the wall to form a narrow passageway.

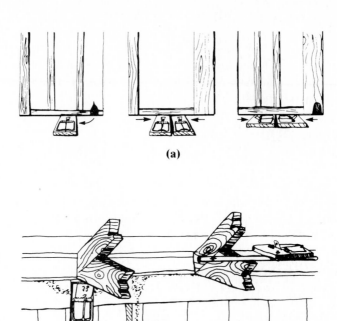

(a)

(b)

Figure 5.6. Effective placement of (a) floor traps: (left) single trap set with trigger next to wall; (center) the double set increases your success; (right) double set placed parallel to the wall with triggers to the outside, (b) overhead traps (particularly useful for roof rats): Trap at left modified by fastening piece of cardboard to expand its trigger size. Traps may be nailed to walls and secured to pipes with wire. (*The Rat: Its Biology and Control.* Division of Agricultural Sciences, University of California, Leaflet 2896, 1981.)

4. Traps set on rafters, beams, or overhead pipes should be fastened firmly to these structural parts so that the dying rodent doesn't drag the trap away and die in some inaccessible wall void.

5. Nail traps vertically where you see greasy swing and rub marks. Do not place traps above food unless the food is covered with a tarpaulin. (See Figure 5.6B.)

6. Place traps for roof rats on tree limbs, under vegetation, on backyard trellises and fences, and other aboveground sites.

Inspect Traps

It may be distasteful to you, but you must inspect the traps regularly to see how well they're working. If they're dirty they won't work as well so, when necessary, scrub them in hot water with detergent and a stiff brush. Rats are put off neither by the odor of humans nor of other dead rodents. You should also move the traps about 2 feet to a new location every few days.

Trapping the Wary Survivors

After most of the rats on your property have been eliminated, it may be hard to trap the survivors. They'll especially avoid traps if they've already sprung one without being injured, so you'll need to use camouflage.

Camouflage Outdoors

Outside, on the ground, sink the trap slightly below ground level and cover its trigger with a small piece of plastic or cloth to keep dirt from clogging the action. Then cover the whole ground area with a layer of fine soil or sawdust. Stones, boxes, or boards can direct an unwary animal along the path to the trap.

Camouflage Indoors

Indoors, bury a trap in a shallow pan of oatmeal, sawdust, or grain. First, however, expose the food for a day or two on the pan without a trap until the rodent readily takes it.

Bait Boxes Protect Children and Pets

To avoid injuring children or pets, set the spring traps in sturdy bait boxes and place them where you know the rodents are running. Such boxes, in various sizes, can be bought commercially. Their entrance holes are just large enough to admit a rat or mouse.

Glue Boards

Glue boards smeared with a sticky substance will catch mice. However, unless two of its feet are caught, a rat can pull itself free. Furthermore, if you're squeamish about disposing of a live victim, its distress cries will disturb you until the animal dies.

Carcass Disposal

In getting rid of dead rats and mice, it's important to avoid any contact with their parasites—fleas, lice, or mites. These are the real carriers of the diseases we can get from rodents. Use a long-handled shovel, long-handled tongs, or two sticks, and wear rubber gloves. If you have none of these, slip your hand into a plastic bag big enough to hold the dead animal, pick up the rodent with that same hand, invert the bag over your catch, and tie the end securely. If you've caught a sizable number, you can burn them if that's legal in your community, or bury them deep enough so that other animals can't dig them up. Should you only catch one or two, you can discard them with the garbage.

Live Rodent Panic

Occasionally a rodent will be caught but not killed by a trap. If this happens to you:

1. Allow yourself one gasp.
2. Get out the dustpan and brush and sweep trap and victim into an empty paper bag. Fold the top of the bag down as far as it will go.
3. Submerge the bag in a bucket of water and hold it down with a brick, a flowerpot or a good-sized rock.
5. Or set the paper bag in a larger, clean plastic bag and put it in the deep freeze. After 2 or 3 hours, move it to a refuse can.
6. You can treat a mouse found in a cereal box in the same way, putting the whole box into the plastic bag and either drowning or freezing it.

Both of these methods are more humane than putting the live animal into the trash barrel to be crushed in the garbage truck's compactor or buried alive in a sanitary landfill.

If you're too squeamish for either of these, recruit a kindly, stout-hearted neighbor for the task.

Shut Them Out

Once you've cut off rodent food supplies and destroyed their harborages, you can concentrate on barricading your home against them. Pest control professionals call this rodent-proofing. To make your home inaccessible to them, you need to know what rodents can do. They are fantastic acrobats.

Rats and Mice Can:

- Go through very small openings; a rat needs slightly more than $\frac{1}{2}$ inch space, a mouse only $\frac{1}{4}$ inch to gain entry.
- Walk along thin horizontal wires or ropes and climb vertical ones. A mouse can travel upside down on mesh hardware cloth.
- Climb the outside of a vertical pipe 3 inches in diameter.
- Climb the inside of a vertical pipe $1\frac{1}{2}$ to 4 inches in diameter.
- Climb *any* pipe or conduit within 3 inches of a wall.
- Climb up a smooth surface if there's a pipe nearby.
- Climb a stucco or brick wall.
- Crawl along any horizontal pipe or conduit.
- Drop 50 feet without being killed. Mice have fallen from a fourth floor without injury. Both have an excellent sense of balance.
- Travel on extremely narrow ledges.
- Gnaw through wood, paper board, cloth sacks, lead pipes, cinder blocks, asbestos, aluminum, adobe brick, and concrete before it's fully hardened.

Rats Can:

- Leap up 3 feet from any flat surface.
- Jump 4 feet across a flat surface.
- Jump horizontally at least 8 feet out from a 15-foot height.
- Reach out about 13 inches.

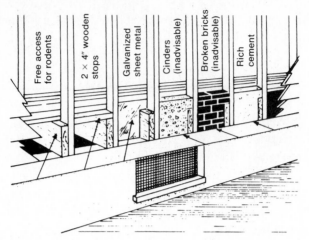

Figure 5.7. Methods of excluding rats from double walls.
(*Rodent-borne Disease Control through Rodent Stoppage*. Atlanta: Centers for Disease Control, 1976.)

Both rats and mice are strong swimmers, even the roof rat and the mouse, which prefer drier environments. All have poor vision but all their other senses are keen.

How can one keep such phenomenal athletes out of a building? By closing off all possible entries and routes of access.

Rodent-Proofing New Construction

If you're remodeling your home or building a new one: Have your contractor nail galvanized sheet metal cut to fit between studs along the sill. Or, fill these areas with a good grade of rich cement. (See Figure 5.7, *c* and *f*.)

A Curtain Wall

Since rats burrow perpendicularly to a wall and do not try to go around it, a curtain wall extending downward and with a horizontal lip stretching out about 12 inches from the base prevents both rat infestations inside and their underground burrowing that can cause walls to collapse. For a home on pilings, a curtain wall of ¼-inch wire mesh will bar rodents (See Figure 5.8.) Another type of rodent stop, illustrated in Figure 5.9, is suitable for a house on a concrete foundation.

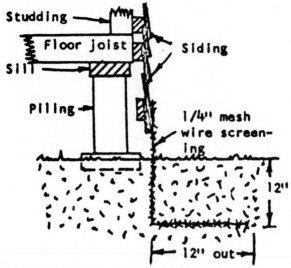

**Use 1/4" mesh wire screening around
the bottom part of the house tightly fastened to the siding and buried
12 inches deep and 12 inches out
from base of building, to prevent
rats from burrowing underground.**

Figure 5.8. Wire mesh rat stopper for house without a
foundation. (*"How to Control Rats on Your Property."*
Pasadena, California: Health Department leaflet.)

Rodent-Proof Materials

Use steel and cement construction wherever possible, with a concrete
foundation and basement floor. Be careful to avoid partially excavated
cellars, and any other unexcavated areas under the building and deck.
Drains and ventilators should be protected with heavy-gauge screening.
Fireplace flues should fit snugly and be kept closed when not in use.

Tight Doors and Windows

All doors and windows should fit snugly, be flashed with metal; screen
doors should close automatically within 3 or 4 seconds. Hardware cloth
covering basement windows should be of ¼-inch mesh.

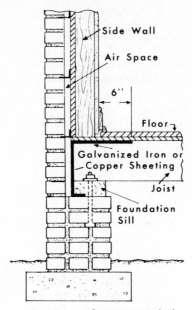

**One type of recommended
rat- proofing between
floor and foundation
in residential construction.**

Figure 5.9. One type of recom-
mended rat-proofing between floor
and foundation in residential con-
struction. (Harold G. Scott and
Margery Borom. *Rodent-Borne
Disease Control through Rodent
Stoppage*. Atlanta: Centers for
Disease Control, 1976.)

Rodent-Proofing an Existing House

If you're already living in the house and have rodents, inspect your
property carefully, starting outside.

Rat Burrows

Check your yard for rat burrows. A well-meaning neighbor may suggest that you gas any that you find. *This is a job for a professional pest
control operator.* The gases used are lethal and, since burrows often
extend under a house, can poison its human occupants. There are two

safe steps you can take to destroy rodents in their burrows:

- Plug all holes, including bolt holes, with cement mixed with broken glass, or
- If your soil is a tight clay that holds water, flood the burrows. This either drowns the rodents or drives them out so they can be clubbed to death.

Barricade Your Roof

Getting to your roof is no problem to rats, whose runways have been traced as high as eleven stories. Remove any vines growing against the building, and prune away all overhanging tree limbs at least 8 feet from the building. Whether for kitchen fans or plumbing, all roof vents should be ratproofed with metal screening as in Figure 5.10.

Pipes and Wires

All pipes and wires coming into your house should be fitted with metal guards. Thin conduits can be blocked with any of the types of guards (A, B, C, D) shown in Figure 5.11. Cone and barrel guards (E and F) are suitable for larger installations. Nails should be set flush with the surface of the barrier so that no rodent can get a toehold on them. They should be far enough apart so as not to serve as a rat ladder (remember, a rat has a 13-inch reach).

Repair Cracks and Holes

See that every crack and hole along the foundation or under the roof is completely patched with masonry. Seal all openings around pipe and wire entries or cover them with sheet metal cut to fit, as in Figure 5.12. If you use cement, mix broken glass with it to keep rats from gnawing through it before it sets. Coarse steel wool can serve as a temporary closure but since it soon rusts, don't depend on it for very long.

Basement Windows

These should be protected with 17-gauge, $\frac{1}{2}$-inch hardware cloth set in removable frames. If you wish, fit 16- by 20-mesh fly screening behind the hardware cloth.

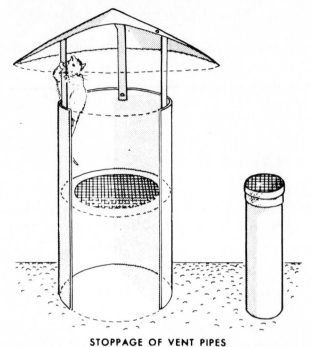

STOPPAGE OF VENT PIPES

Figure 5.10. Stoppage of vent pipes. (Harold G. Scott and Margery Borom. *Rodent-Borne Disease Control through Rodent Stoppage.* Atlanta: Centers for Disease Control, 1976.)

Attic Vents

These should be firmly covered with ¼-inch mesh hardware cloth.

Letter Drops

A letter drop less than 18 inches off the floor should have an opening smaller than ½ inch. If the opening is wider, install a spring-closing cover on it.

Floor Drains

Floor drains in your garage, basement, or at your pool should be sturdily screened.

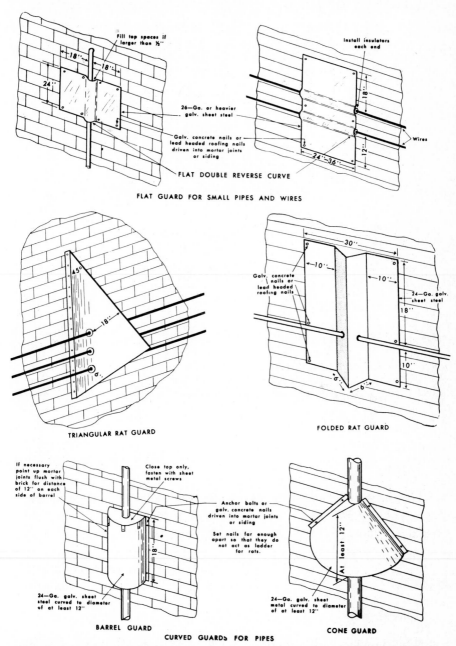

Figure 5.11 A–E. Types of rat guards for pipes and conduits. (Harold G. Scott and Margery Borom. *Rodent-Borne Disease Control through Rodent Stoppage*. Atlanta: Centers for Disease Control, 1976.)

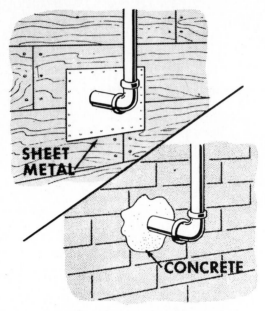

STOPPAGE OF OPENINGS AROUND PIPES

Figure 5.12. Stoppage of openings around pipes.
(Harold G. Scott and Margery Borom. *Rodent-
Borne Disease Control through Rodent Stoppage*. Atlanta: Centers for Disease Control,
1976.)

Doors

Protect the edges of wooden outside doors with metal channels and
cuffs, as in Figure 5.13, to make it impossible for rodents to gnaw through
them. All doors to the outside, including basement doors, should have
metal flashing.

Thresholds

While looking over your doors, check their thresholds. If they're worn
or broken, replace them or cover them with 24-gauge galvanized metal.
The distance between the bottom of the door and the threshold should not
be more than $\frac{1}{2}$-inch; $\frac{3}{8}$-inch would be better.

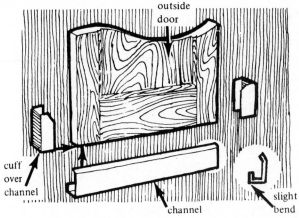

Figure 5.13. Cuff and channel rodent stoppage for wooden doors. (*Control of Domestic Rats and Mice*. Atlanta Centers for Disease Control. Homestudy Course 3013-G, Manual 11, 1982.)

Rodent-Proof Materials

In modifying your home, wherever you can, use rodent-proof materials such as 24- to 26-gauge sheet metal, expanded metal 28-gauge or heavier with mesh no larger than $\frac{1}{4}$-inch, iron grilles of equal gauge with slots $\frac{1}{4}$-inch or less wide, and perforated metal with $\frac{1}{4}$-inch holes. Other materials rodents can't penetrate are concrete, brick, mortar, and tile.

The Catch in Rodent-Proofing

In some ways rodent-proofing is a catch-22 situation. When a building has been rat-proofed, its potential for harboring mice is increased because fewer animals are competing for food and living space. If your visiting rodents are mice, you'll have to be more meticulous about sealing the smallest openings. Most city building codes, if conscientiously followed, will effectively rat-proof a structure, but they rarely completely mouse-proof it. Thorough sanitation is your best defense against mice. Once their nests and food supplies are eliminated, mice may move away of their own accord.

Keeping the Rodents Out

To keep your home rodent-free, check the structure at least once a year. Constant settling of the ground and drying of the wood can open new entryways, while any changes in plumbing lines or electrical wiring can also give these little mammals easy access. Here's a checklist to help you keep all possible points of entry plugged.

Lower Floors

• Are all openings in foundation and under siding sealed?   _____

• Are all basement windows screened as described above?   _____

• Are all floor drains covered with galvanized heavy-gauge strainers?   _____

• Are all toilets in good repair?   _____

• Are all pipes and wires entering the house protected with rat guards?   _____

• Do all screen doors close automatically?   _____

• Does the letter drop have a spring-closed cover?   _____

• Are all openings around pipes and wires entering the house sealed?   _____

Upper Floors

• Are tree limbs pruned at least 8 feet away from the house?   _____

• Are attic vents covered with sturdy hardware cloth?   _____

• Are all openings and intersections at the roofline sealed?   _____

Note: If someone offers you a yellow paste resembling peanut butter and guarantees it will kill rats and mice, refuse it. It's yellow phosphorus, a poison so powerful it can kill a small child who walks over it barefoot and *there is no antidote for it.* It may be called J-O Paste, Paste Electrica, Rough-on-Rats, or Stearns Electric Paste. Illegal for use as a pesticide in the United States, yellow phosphorus comes into this country via Latin America.

Control a Public Responsibility

You may be the world's most meticulous householder but if you live in a community with rat havens like open dumps and hog-fattening plants, you're going to be coping with rodents.

Corral some concerned neighbors and start to prod city hall with newspaper and television messages, door-to-door leaflets, and school programs. Don't let up the pressure until authorities have replaced the dump with a sanitary landfill and the pig-feeding station has been cleaned up.

Dead Rodent Odor

If a rodent has died in your wall voids, you'll have to deal with the odor sooner or later. You might consider taking a long vacation, but there are less expensive ways to solve this problem.

Fans, including the vent fan in a room air-conditioner, help drive out the stench. You can also use an aerosol spray or mist, or a cotton wick dipped into a bowl of one of the following:

Bactine

Dutrol

Isobornyl acetate

Neutroleum alpha

Quarternary ammonium compounds

Styamine 1622

Zephiran chloride

Oils of pine, peppermint or wintergreen, formalin, anise and activated charcoal are also effective odor-masks.

If you can locate the carcass inside the wall (Do flies gather there? Does your dog keep sniffing there?), drill a small hole through the wall a few inches above the floor and as close to the odor source as possible. Using a narrow tube, pour in one of the masking agents listed above. This is usually the fastest way to clear out a stench.

Why Not Just Call in the Cats?

Wouldn't it be simpler just to get a cat and turn it loose? Not really. Professional pest control operators think a cat is worthless against an infestation of rats. Says one New York City rat control officer, "You put a cat in a cellar with a heavy rat infestation and you got yourself one very dead cat, believe me.''

Some dogs—a silky terrier or a toy fox terrier, for instance—may be good rat catchers, but generally a cat or dog will get the worst of it in any dustup with a ferocious buck rat. In packs, rats will gang up and best a wild boar, even a tiger.

Studies have shown that in residential areas, cats kill only about 20 percent of the number of rats that must die every year to maintain a stable rodent population. On farms, cats kill enough rats to prevent an upsurge in rodent population. In neither location do cats exterminate the rats.

The fact is that cats and dogs may actually draw rats to an area. A rat can live very comfortably under a dog house, eating and drinking the pet's food and water while the rightful recipient is asleep. In general, cats and dogs just keep the rodents out of sight, where the pets can't reach them. However, pets may be useful in keeping an area free of rodents *after* the vermin have been killed by other means.

And don't count on owls and hawks to do the job of rodent eradication either. Along with foxes, snakes, humans, cats, dogs, and parasites, rodent predators cut down the pest population only temporarily. The weight of scientific evidence is that the number of *predators* is actually regulated by the number of their prey, and not the other way round.

Electronic Repellents

In the past few decades we've come to think of electronics as a kind of magic wand that makes all things possible. Serious efforts have been made to apply electronic principles to rodent control. For a while electromagnetic devices were marketed that claimed to drive off rats and mice. They soon proved worthless, however, and the Environmental Protection Agency has stopped their sale, charging that their advertising was false and misleading.

Ultrasonic devices seemed to make more sense. Rats and mice have intensely acute hearing. Any loud noise sends them bolting for escape. Environmentalists and food processors once thought such devices held great promise for rodent control. Sad to say, that promise has not been fulfilled. Ultrasound is directional, does not penetrate stored objects, and loses intensity very quickly with distance. Tests run by the Division of Agricultural Sciences of the University of California have all proved negative.

The rodents themselves have passed the ultimate judgment on ultrasonic repellers. California state health official Minoo Madon found a family of mice nesting in one!

Odor Repellents

Certain odors are known to repel rats and mice. Camphor gum and spurge (*Euphorbia lathyris*) hold off mice, as catnip and dog fennel are said to do. Unfortunately, the effect of these botanicals is only short-lived. They will probably not succeed where there are established rodent colonies. Managers of food warehouses and processing plants would be elated if certain odors or electronic devices could drive rodents permanently from their facilities. So far none have proved to be of much help.

Future Controls

Environmental researchers are working along several tracks to sharpen our weapons against domestic rodents. One route being studied is controlling rodent fertility. Another is lacing baited poisons with pheromones (sex lures) to make them more attractive to these wary animals. A third is biosonics—that is, mimickings or recordings of rodent distress sounds, played back with good acoustics and at low intensity to lure the animals into traps. Practical use of any of these is most likely years down the road.

Until something is proven to be better, we'll have to continue to depend on sanitation, rodent-proofing our buildings, and well-laid traps to keep rodents—adaptable, wily, prolific—from overrunning our homes and farms.

CHAPTER 6

The Fearsome Fly

Once, a long time ago, a tailor sewing in the window of his shop was bothered by a swarm of flies. Whacking about with a piece of cloth, he killed seven of the buzzers at one blow. So proud was he of this feat that he stitched his belt with the words, "Seven at One Blow," and left his shop forever to seek his fortune. Parlaying his skill as a fly slayer into a reputation for great bravery, he won the hand of a princess along with her father's kingdom.

The plucky tailor well deserved his good fortune for he had destroyed seven of mankind's most dangerous enemies.

Dangers from Flies

Along with their cousins the mosquitoes, flies have caused more human and animal deaths than any other insects. Although the evidence in some cases is circumstantial, the common housefly, one of several kinds of filth flies, is the probable carrier of more than sixty-five human and animal diseases, according to entomologist Walter Ebeling of the University of California. It has been linked to outbreaks of cholera, anthrax, dysentery, tuberculosis, typhoid fever, plague, yaws, leprosy, tapeworm, gonorrhea, and polio. According to The Centers for Disease Control, the housefly is a greater danger to humans than any other species because of its close association with us, its filthy habits, and its ability to transmit germs.

Not all flies are dangerous and some are even helpful. Cluster flies swarming on a warm winter day are a harmless nuisance since their main breeding environment is earthworm-rich soil. Hover flies pollinate useful plants and eat destructive aphids. Tachinid flies lay their eggs in pest caterpillars, which then serve as food for the tachinid larvae. The larva of the black garbage fly, which is not found near human dwellings, can kill up to twenty housefly young a day.

However, blowflies, stable flies, false stable flies, houseflies, and little houseflies should be kept as far from our homes as possible (See Figure 6.1.). Blowflies, either bluebottle or greenbottle, that buzz around a lamp at night or settle on a kitchen counter can fly up to 28 miles from their breeding grounds in garbage dumps or slaughterhouses to contaminate our food. They often lay their eggs on fresh meat minutes after it's exposed. If swallowed, the eggs and/or larvae can make us very sick.

Gnats, or midges and fruit flies—both true flies—are also a danger. So tiny they can pass through ordinary window screening, some of these minicreatures give a painful, long-lasting bite. Some are bloodsuckers. Entomologists have traced outbreaks of conjunctivitis (pinkeye) and trachoma to the eye gnat, which ruptures the eye's protective membrane. Untreated trachoma leads to blindness. Fruit flies, breeding in rotting fruit, uncooked food, and excrement, are also most unsavory visitors.

Fly Problem Growing Worse

Not long ago health authorities felt that the fly problem in the United States was not serious, but now they're changing their minds. For the last century our population has been encroaching deeper and deeper into the countryside. At the same time, as our numbers have grown, we consume vastly more meat and poultry, which have to be produced somewhere.

Cattle, hogs, and chickens, instead of being raised as they once were on widely scattered farms where the insects drawn to their droppings were soon eaten by predators, are now gathered in mass feeding facilities, where important insect eaters have been eliminated. Unless meticulously managed, these "meat factories" quickly become "pest factories."

Your new suburban home may be only a short distance from a cattle feedlot or poultry ranch. In addition, the tender new lawn you're conscientiously fertilizing will add to your fly problem. You may be coping with flies for several years, until neighborhood lawns are well established and additional development pushes the livestock further away.

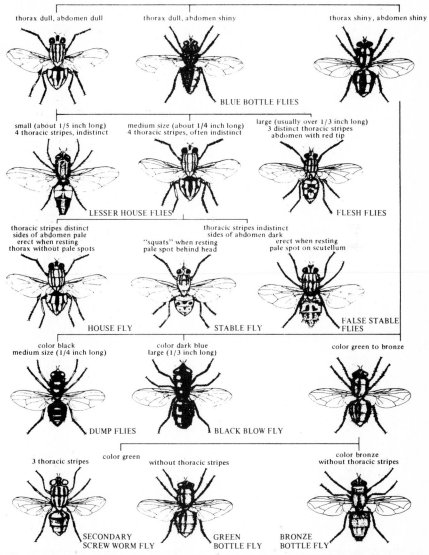

Figure 6.1. Pictorial key to common domestic flies. (*Flies of Public Health Importance and Their Control*. Atlanta: Centers for Disease Control. Homestudy Course 3013-6, Manual 5, 1982.)

Homeowners who've only recently moved to a new development from a city or older suburb may find the flies worse than do people who've lived in the area for a while. Where countryside and farm are giving way to suburb, there will be more flies, rodents, snakes, and wasps than you're used to—they come with the territory. Until public sanitation is well established, you'll need to have more tolerance for small wildlife while rigorously monitoring the sanitation around your own home.

Nature of Flies

What Makes Flies Dangerous

Since many species of flies feed and lay their eggs in cesspools, stagnant water, dung heaps, garbage cans, and any rotting animal or vegetable matter, bacteria enter their bodies and are also trapped on their leg hairs. Their larvae, or maggots, feed on the filth from 4 days to 2 weeks. After pupating (resting in an outer case) for less than a week, a fully developed adult fly emerges. Incidentally, small houseflies are not the pest's young; they are members of another species just as foul.

When they get into our homes, sample our food, walk on our dishes, or sip our sweat, houseflies contaminate us with their eggs, larvae, bacteria, and excreta. Outbreaks of diarrhea have been found to correlate with the warm weather surge in the fly population. So don't shrug off those black buzzers hovering in your entranceway or around your patio.

Although chemists have developed a whole battery of powerful sprays to control pest flies, in a few short generations (there may be ten generations a summer) the species often becomes immune to the chemicals.

Starve Them Out

What You Can Do

The key to fly control can be summed up in one word—sanitation. Every adult fly near your home represents a breakdown of sanitation procedures. To a great extent we make our own fly problem by our careless handling of food waste and other decaying organic matter. Some years ago two scientists undertook to count the number of flies produced by inadequate garbage cans. These unsung heroes found that in summer the average barrel bred over 1000 flies a week. One barrel, which must have belonged to a dedicated slob, produced nearly 20,000 in one week.

How you handle food waste, then, is critical. Like most insects, flies have a keen sense of smell. Though drawn by almost any food odor, they don't just go for the pleasant ones; a stench that turns your stomach sets this pest's mouth watering. One of your major goals in handling food waste must be to control the odors of decay. The following simple steps, made your habit, will do just that.

Indoor Sanitation

1. Rinse clean all cans, jars, dairy cartons, and meat or poultry wrappings and drain them well before throwing them out.
2. Keep a plastic bag in your kitchen garbage pail for food waste. Drain the scraps well, and twist the bag closed each time you've put refuse into it. At the end of the day close the bag with a twist tie or rubber band and take it outside—to the incinerator if you live in an apartment, to a larger, closely covered garbage barrel if you live in a single-family house.
3. Handle throwaway diapers hygienically. Unless they're disposed of carefully, those cushiony little pants can turn a family's garbage cans into disease-breeding toilets.

Outdoor Sanitation

Garbage Barrels

Garbage barrels outside should be in good repair and properly stored.

Lids. It does no good to set garbage outside if the barrel's lid is missing or even just cracked. A fly can squeeze through a slit to get at a moist, decaying meal. If you can't replace a cracked lid with a snug one (some dealers sell lids alone), get a new container. Tight covers on sturdy barrels can reduce the local fly population by 90 percent if garbage is well wrapped.

Container Storage. Metal barrels should have recessed bottoms to prevent rusting. All should be stored on a sanitary rack at least 18 inches off the ground to minimize bacteria-favoring moisture underneath. (See Figures 5.3–5.5.)

Never use trash bags alone as garbage receptacles. Cats, dogs, rats, and—in the far west—coyotes, can quickly tear them open.

Patio and Garden

After a barbecue, be sure to include the grill in your cleanup so that bits of meat or fish don't give flies their own cookout. Keep dropped fruit picked up.

Lawn Maintenance

Anyone who's ever fertilized a lawn or been in the neighborhood when manure is being spread knows how flies swarm to it, but even without fertilizer a lawn can be a fly source. Decomposing trimmings may also add to your fly problem. Not long ago an entomologist raised twenty-five filth flies from grass clippings caked on the blade guard of a power mower. To prevent lawn-bred flies

- Spread manure in a thin layer so that eggs and maggots will die of heat, cold, and/or dryness.
- Gather lawn clippings in a plastic bag and put it out for the collectors.

Dogs' and Cats' Droppings

A fine dog adds much to life, but unless you dispose of its droppings promptly and properly, the local pest population will soon outrun your ability to curb it. The same is true of that rare being, a sloppy cat. And you'll not only be coping with more flies but with a lively bunch of cockroaches too.

There are several sanitary ways to deal with pet droppings:

1. Bury droppings. Run a sprinkler for several hours in an unpaved part of your yard, and then dig a hole 2 or 3 feet deep. Drop in each day's cleanup and cover it with a light layer of the excavated dirt. Be sure, however, that your dog is free of worms as the eggs can live a long time in the ground and be picked up by anyone walking there barefoot.

2. Dog latrine. Set a commercially available dog latrine in the ground. Properly maintained this should work if your dog is not too large.

3. Dispose of droppings in a plastic bag and then set it in the garbage barrel.

4. Flush them down the toilet.

5. When you can't dispose of the piles right away, sprinkle them with a thin layer of sand or sawdust to minimize stench and moisture.

6. Confine your pet. Thoughtful people don't let their animals set up fly breeders on the neighbors' lawns or public property. In New York City this became such a serious problem in 1981 that sanitation police began a crackdown on people whose animals filthied the city's streets.

Dog bones will also pull in flies. After your pet's had enough gnawing, wrap the bone in plastic and throw it out. Incidentally, bones left outdoors attract rats.

Neighborhood Sanitation

So you're always meticulous but there are still too many flies in your area. What can you do? Check for rotting organic matter nearby or stagnant water. If the problem is severe, call your property owners' association or health department. They can contact careless individuals or alert the whole neighborhood to the potential for disease. Officials can also have standing water on public property drained or order empty lots being used as dumps or pet latrines to be cleaned up and fenced off.

Wipe Them Out

Traps, Swatters, Light

As long as people generate waste, flies will be with us. Your goal is to keep them to a minimum. An outdoor fly trap can do a great job of fly control.

1. A simple one, *Fli-Lur,* currently selling for less than $5, looks like two small tube-cake pans set edge to edge. It is baited with a bit of meat, stale beer, or sugar and then half filled with water. If you use meat the trap will smell in a few days but that soon fades. Set it at a distance from the house. The flies crawl through a grid to get the bait, and then they drown. Check the trap twice weekly to see that it hasn't dried out. A dry trap won't catch any flies.

    Corinne Ray, director of the Los Angeles Poison Information Center and owner of two horses, keeps a Fli-Lur trap on the roof of her stable. Others are over her patio and kitchen. Despite the horses, she has few flies. When asked who empties the traps, she replies with a smile, ''My husband.''

2. A less expensive trap is a lidlike strainer that fits over a jar. Be sure to set this trap where it can't shatter.

3. A suggestion from *Mother Earth News:* Shape a piece of paper into a cone and insert it into the neck of a baited jar.

4. You can still buy flypaper, but some public health authorities consider it fairly ineffective.

5. For the occasional invader, an expertly wielded fly swatter is as potent as any aerosol spray and a lot safer for you. A quick sudsing in hot water after the kill, followed by sun or air drying, takes care of any contamination on the swatter. If you can get the hang of using a rolled-up newspaper or magazine for a swatter (I never have), you've got the ideal weapon—effective, sanitary, disposable.

6. Like many winged insects, flies are drawn to light. To get rid of them in the daytime, darken the room they're in and open the door to the light, the brighter the better. Open the screen door and they'll soon fly out. If the screen door is closed, the flies are sitting ducks. If you're not too good with a swatter, spray the insects with rubbing alcohol for a quick kill and few escapes.

7. If they get in at night, try this: Before you go to bed, close all the drapes and window shades, leaving a narrow slit at one of them. In the morning, the flies, torpid in the cool air and dozing on the bare, bright strip of glass, will be easy prey.

Nature's Fly Traps

Maybe you've had occasion to use a country privy. Were you surprised that despite the fragrance there were few if any flies? Spider webs veiling the window and wasp nests in high corners were signs that the local sanitation crew was on the job.

Spiders

Flies soon become immune to the chemist's most powerful toxins, but in millions of years on earth no fly has ever developed immunity to any spider. You may be even less fond of spiders than you are of flies, but most of the small spinners are harmless (see Chapter 13 for control of spiders). The species as a whole, even the poisonous ones, rarely bite.

Although spiders prey on various insect pests, their favorite dish seems to be flies. So when a harmless spinner sets up housekeeping in an out-of-the-way corner of your property, try to resist the urge to sweep away the web and step on its builder. Chances are it's no danger to you. Your tolerance will be well repaid for each of these skillful weavers can destroy up to 2000 pests in its short life.

If your washing machine is under a window in your otherwise dark garage or basement, you may regularly find webs at the window. Let them stay. Drawn by the light and moisture of the laundering, flies buzz against the window, where a watchful spider soon adds them to its larder. When the web gets too large, brush half of it away. In a matter of weeks the spider or its children will rebuild it and continue their patrol.

Note: To be certain that the spider working your territory is not a potential danger, put on a pair of heavy work gloves, scoop the animal into a small box or jar, and take it to your local arboretum or community's college biology department for identification.

Wasps

A more temperamental but possibly even more effective ally in controlling flies is the wasp. Of course one doesn't want to get too chummy with a wasp or let it indoors. However, it's a voracious pest predator, feeding its young on many different kinds of insects. A century ago, iron-nerved Americans hung wasps' nests near their stable doors to keep down flies. A colony of yellow jackets or paper wasps nesting in a little-used corner of your yard will provide more thorough and longer-lasting pest control than any chemical.

Shut Them Out

Flyproofing a house is fairly easy and most of the controls are mechanical.

Screens and Fans

Tightly fitting window screens (16-mesh is the standard) and outward opening, self-closing screen doors are as foolproof a fly barricade as anyone has yet invented. Whether metal or plastic, the screening must be in good repair and firmly secured to its frame.

If your door is used a great deal, install a ceiling fan above it that moves at least 1500 cubic feet of air downward every minute. This is a ploy you'll find in many butcher shops and restaurants, where doors are opened as much as they're closed. If you have a covered patio and enjoy outdoor meals, set an overhead fan there too.

Repellents

Are you interested in experimenting with herbal controls? Try growing tansy near your kitchen door or wherever flies tend to cluster. Though it's not a substitute for meticulous sanitation, rural people once found the plant a good fly repellent.

If you have the space, plant a camphor tree near your kitchen, recommends Dr. C. E. Schreck of the United States Department of Agriculture's Insect Repellent and Attractant Project. Most insects shy away from the tree's pungent aroma.

Other effective fly repellents are oil of cloves and pine oil. Deet, commonly used to keep off mosquitoes and ticks, will deter this pest as well.

Pest Flies That Need Different Controls

Several pest fly species pose health dangers and have habits different from those of houseflies. You'll need to adopt other strategies in controlling them.

Vinegar or Fruit Flies

This tiny fly's ability to reproduce quickly makes it an important tool in genetic research. It also makes it a potentially serious pest. If you make your own wine or beer, you'll have to be meticulous about keeping the containers, lids, and utensils clean inside and out.

Habits: Feeds and breeds on any fermenting substance. Easily penetrates standard screening. Common attractants around home are:

- Containers of cider, vinegar, beer, wine whose contents have spilled over
- Ripe fruits and vegetables
- Juices, ketchup, sour milk

- Souring scrub water and mops
- Dirty garbage cans

Controls: Control should aim at eliminating breeding and feeding sites and denying fruit flies entry.

- Consider changing to a finer-meshed house screen.
- Clean up any fruit peelings, spilled juice, or milk under refrigerator or stove.
- Store ketchup in refrigerator.
- Rinse thoroughly all dairy cartons, juice cans, and ketchup bottles before discarding.
- Don't let scrub water stand in floor crevices.
- After washing floor, dry it with a dry rag.
- Set wet mop, thoroughly rinsed, in sun for a few hours.
- Use unshaded window slit to trap them (see point 6 under "Traps and Swatters").

Drain Flies

If a small, brownish-gray, hairy monster emerges from your bathroom drain, don't panic. You're not in a Japanese horror film gone Lilliputian. It is a drain fly and, depending on where it's been living, may or may not be dangerous. Since you probably find the whole idea of flies in the plumbing revolting, you'll want to wipe out any of the same ilk remaining in the pipes. *Do not pour insecticide down the drain.* It's illegal in many communities and hazardous to everyone in the vicinity.

Habits: Prefers gelatinous accumulations in sewers, septic tanks, tree holes, and rain barrels. Also likes damp, dirty garbage pails and moist birds' nests.

Controls: Eliminate breeding and feeding places.

- Close up any nearby tree holes.
- Discourage nest-building birds.
- Keep rain barrels covered in dry weather.
- Drain all standing water.
- Handle garbage hygienically (see "Outdoor Sanitation").
- Occasionally scour insides of drains with a swab of paper wrapped in waxed paper.

- About once a month pour a cup of chlorine bleach down each drain (*not* the garbage disposal) and let it stand a few hours to prevent buildup of goo.

The Little Housefly

This pest appears earlier in the spring than the housefly and tends to become inactive or even disappear in hot weather.

Habits: Likes animal droppings and other decaying organic matter. Also likes honeydew, the syrupy substance secreted by aphids and other garden insects. The syrup eventually molds and becomes sooty black.

Controls: Eliminate breeding and feeding places.

- Monitor food waste carefully.
- Eliminate honeydew-producing garden pests. These include aphids, mealybugs, scale insects, and whitefly. (See Chapter 12 on garden pests.)

Midges and Gnats

As more people move out to undeveloped lands of the desert, mountains, and seashore, midges and gnats, which usually annoy us only when we're outdoors, can be a serious problem. Biting midges, eye gnats, buffalo gnats, turkey gnats, and valley black flies are only a few of these troublesome species.

Some can cause serious illness, including swelling of an entire arm or leg, and fever. When they swarm against an automobile radiator, the engine can overheat. Massed on a windshield, they cause accidents. They are such a nuisance in Florida and the Caribbean that they've caused dropoffs in tourism.

Because these flies' breeding places are so vast and diverse—lakes, streams, irrigation canals and ditches, stagnant water, tree sap, and marshes—health officials applying suitable insecticides are the best ones to control gnats and midges.

The most you can do in areas where they're a serious nuisance is to install very fine screening material such as the bolting cloth used in grain mills. When these insects swarm, stay indoors with the windows closed. When you must go out, use a repellent such as deet (diethyl toluamide).

If you're planning a vacation in the mountains or desert or at the seashore, ask health authorities in the region you'll be visiting when gnats and midges are usually active. Then plan accordingly.

Community Sanitary Standards Recommended by the Centers for Disease Control

Adequate community sanitation and refuse control are essential to any program aimed at eliminating flies. What does "adequate" mean here? How does your community measure up? What happens to refuse after it leaves your control? How can you know what good community sanitation is?

The following standards have been established by the Federal Centers for Disease Control for safe refuse handling. These recommendations should be followed closely by every community official concerned with citizen health.

Privies and Septic Tanks

There are still places in the United States without indoor flush toilets. Many are in substandard rural and semirural homes or they are public facilities in poorly developed small communities or campgrounds. Properly installed and maintained septic tanks can be a preliminary measure in places far from a modern sewage system, but they are acceptable only temporarily. As soon as it can be arranged, the household or public recreation area using a cesspool should be hooked into an up-to-date municipal sewerage network.

For households or recreation facilities in remote areas, composting toilets may be the best long-term solution. There are several types currently on the market; they're designed to be odor free and produce usable fertilizer. Check with your local health authorities to see which is acceptable in your area.

Industrial Waste

City and county health and sanitation officials should continuously monitor any slaughterhouses and packing plants, canneries, and feed mills within their jurisdiction. Such facilities can be prime sources of disease-bearing pests.

Refuse Collection

"The collection system," say the Centers for Disease Control, "must be designed for the improvement of sanitation and not for the convenience of the collection agency."

Refuse should be collected from residences twice a week in summer, daily from restaurants, hotels, and food markets. Collectors should be careful not to spill garbage or damage the containers. Their trucks should be closed, of the packer type, and kept clean.

Final Disposal

Too many towns and cities in the United States today still drop their garbage on vacant acreage at what officials think is a safe distance from the community. So great is this problem that the Environmental Protection Agency has launched "Mission 5000," designed to eliminate 5000 open dumps in this country.

If you think you're safe because your town burns its refuse at a dump "way out" in the country, you're wrong. Houseflies have been found 100 miles out at sea, blown there by the wind.

A burning dump is a menace to a community, attracting filth insects and verminous rodents. In addition, exploding glass and aerosol containers can injure anyone in the vicinity. Proper disposal of a city's garbage can be done in any of three ways: individual garbage disposals for each household, incinerators, and sanitary landfills.

Garbage Disposals

When you use a garbage disposal, your pulverized food waste becomes part of the city's sewage, and is treated at its reclamation plant. To prevent the occasional stuffed sink (Why does this always happen during a dinner party?), make sure all the garbage is completely flushed away by running a strong stream of water for a full minute after you shut off the appliance.

Incinerators

Incinerators, whether for a whole city or a single apartment building, should operate at temperatures from 1400 to 2000°. A defective installation will merely char the garbage, barbecuing it nicely for the rats and cockroaches.

Sanitary Landfill

In a sanitary landfill, compacted garbage is dumped on swampy or otherwise useless land some distance from the city. There it's covered *every day* with at least 6 inches of earth. When the parcel is completely filled, it's buried under a minimum of 2 feet of soil.

Some cities find it sufficient to shred, mill, or pulverize the garbage to destroy maggots before dumping it, and use no soil cover.

As a side benefit, a sanitary landfill reclaims otherwise unusable land and makes it habitable.

If public health officials, elected or appointed, are permitting any or all of the above-mentioned fly breeders to menace your community's health, become a fly yourself—a gadfly—and nag them until the hazards are eliminated.

Biological Control—A Promising Approach

Despite the many species of pest flies, the one of most concern remains the common housefly. Scientists continue to seek ways to keep its numbers within safer limits. Absolute safety would be no flies at all, but this is impossible. Since chemicals are clearly not the answer to fly eradication, researchers are looking into biological controls.

Nonpest fly larvae that compete with housefly young for dung, and larval parasites that destroy hatchlings are under study. Both seem to hold promise. The release of sterile male flies has had spectacular success east of the Mississippi against the once-devastating screwworm fly. This insect's maggot can infest 20 percent of a cattle herd, kill 20 percent of those infested, and be ingested by humans eating the meat. Infestations in western states will end completely when scientists can figure out how to prevent reintroduction of the screwworm fly from Mexico.

Progress in biological control will take time. Until we learn how to disrupt the housefly's breeding cycle and how to apply those techniques throughout a state or even the nation, rigorous monitoring of animal, human, and food waste in our homes and communities remains our best means of control.

PART THREE

PESTS
OF THE BODY

CHAPTER 7

The Mosquito—
A Deadly Nuisance

Over the centuries, the mosquito has caused humanity more grief than any other insect, more even than the plague-bearing rat flea that killed three-fourths of the people in Europe and Asia 600 years ago. Historians say it was partly responsible for the fall of Rome, infecting many of the city's inhabitants with malaria.

Almost singlehandedly this dainty insect brought the construction of the Panama Canal to a complete halt as yellow fever or malaria felled its French builders one after the other. A century later the moon shot was imperiled when Louisiana rocket builders threatened to quit because day-biting mosquito hordes were attacking their children. Magnificent Yosemite National Park was uninhabitable until health authorities cleared out the area's dense clouds of mosquitoes.

In all fifty states, most efforts against public pests—80 to 90 percent—are aimed at suppressing this insect. Malaria, yellow fever, dengue fever, and encephalitis (brain fever) are all mosquito-borne diseases. Even our dogs don't escape for mosquitoes give them heartworm.

Cost of the Mosquito

Money spent on suppressing the mosquito is astronomical. The cost of materials that kill adults or larvae, and the wages of workers to apply those materials on public lands; reduced cattle weight and less milk, even cattle deaths when swarms are heavy; drop-offs in tourism; and poor labor efficiency are all part of the tribute we pay the mosquito. That high whine and the silent bite that follows can ruin much more than an evening outdoors. Not only are communicable diseases a threat, but people can also go into shock from bites or develop secondary infections.

Malaria, Yellow Fever, Dengue Fever, Encephalitis

Malaria once claimed 100,000 victims a year in the United States. Although no longer the danger it once was here, it continues to disable at least 200 million people every year worldwide. Globally it causes more deaths, up to 5 million annually, than any other transmittable disease.

The mosquitoes that carry malaria are found in many parts of this country, but massive public drainage projects and other successful controls of breeding sites have made the disease relatively rare here. However, during World War II, it staged a comeback when over half a million GIs brought it home from overseas. During the Vietnamese war, as many soldiers were sent home in one year because of malaria as were returned wounded.

Today, most cases of malaria in the United States are transmitted by the needles of heroin addicts. Pockets of it also fester among Asian refugees. Should these unfortunate people come into contact with the mosquito that hosts the malaria organism, the possibility of an upsurge in the infection would be grave.

For centuries quinine in various forms has eased malaria's chills and fever. Lately, however, the parasite that causes malaria has been resisting the drug. Making matters worse, the mosquito transmitting the disease (the anopheles) is becoming resistant to the insecticide health workers have depended on to hold down its numbers.

Yellow fever, which raged in this country as late as 1905, is also a gift of the mosquito. Modern water management has practically wiped it out here.

Dengue fever, carried by the same species of mosquito that brings yellow fever, is another serious illness. It's widespread in warm climates. Other than good nursing care, there is no treatment for it.

In the United States the greatest threat to humans from the mosquito is **encephalitis,** an inflammation of the brain. It can be as mild as a slight summer cold or it can be potentially fatal. Victims who are children, if they survive, are often left with seriously damaged nervous systems. There is no medicine that can cure human encephalitis (horses get it, too, but they can be immunized), and the only preventive for us is mosquito control—never-ending, thorough mosquito control.

Encephalitis outbreaks occur every year in the United States, often after a flood. In 1975, one of the worst years, more than 2000 cases were reported.

Nature of the Mosquito

It is almost weightless—10,000 can tip the scales at less than an ounce—a leggy creature with long, narrow wings. Rather dainty, if you can take the time before slapping to look at it carefully. Some are even beautiful, with splashes of silver gleaming on dark bodies. Despite its delicate appearance, the mosquito is without doubt one of the most dangerous animals on earth, carrying deadly diseases to millions of human beings every year.

The mosquito is actually a fly; its name is the Spanish word for "little fly." Depending on which expert you consult, there are 3000 to 4000 different species of the insect, 400 to 500 capable of transmitting diseases. Although both sexes feed on flower nectars, the male never bites. The female, however, must have a meal of blood before she can lay fertile eggs.

Some species feed during the day, some at dusk, others at night. In warm climates they're abundant all year. In cooler regions, they can appear as early as February, when melting snows form the quiet pools that the female seeks to shelter her eggs.

Mosquito's Life Cycle

The mosquito has four life stages—egg, larva, pupa, adult. The egg is laid on or near water, or where water will eventually cover it. The larva and pupa live under water, while the adult lives on dry land. (See Figure 7.1.)

All mosquitoes require still water on which to drop their eggs. Depending on the species, the site may be:

Permanent stagnant water	Lakes, ponds, salt marshes
Temporary still water	Ditches, borrow pits, clogged streams, irrigation canals, containers around homes
Flood waters	

All three kinds of water sites abound in and near cities and out in the countryside. If the eggs are laid on a small puddle that dries up before they hatch, the larvae will emerge almost as soon as water accumulates there again, even if, for some varieties, it's *5 or 6 years* later.

The larval, or wiggler, stage may last a week; then the animal becomes a pupa, or tumbler. Three or four days later, the adult appears, ready to mate and, if female, hungry for a blood meal. Though some species can fly many miles, some of the most common, like the house mosquito, will only travel to the nearest blood feeding.

Tougher than the Insecticides

Pesticides brought disease-carrying mosquitoes pretty much under control for a while. When DDT came on the scene right after World War II, malaria, yellow fever, and encephalitis seemed to fade rapidly to ghostly threats from the past. As early as 1952, however, a major carrier of encephalitis was showing immunity to the pesticide, and health authorities were worried until new chemicals were developed. Throughout the United States, and in other parts of the world, mosquitoes now bounce back from what were once our most dependable chemical weapons. By 1976 California mosquito larvae (the easiest stage to kill since they're not dispersed like the adults) were resisting every available larvicide that does not pose a serious threat to other living creatures.

Wipe Them Out

Water Management: Key to Mosquito Control

So today we're back where mosquito control started a century ago. Our experience with the Panama Canal (a major breakthrough in the

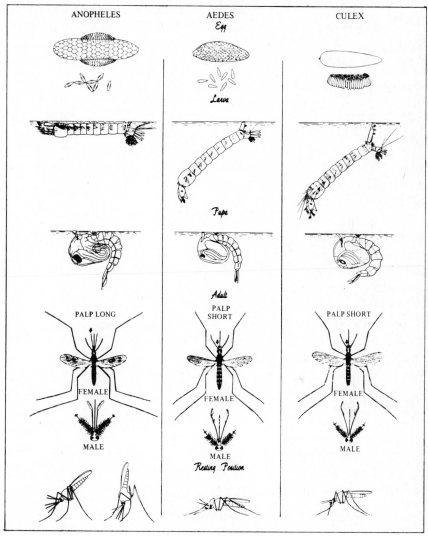

Figure 7.1. Life cycles of three disease-bearing mosquitoes. (*Insect and Rodent Control: Repairs and Utilities.* War Department Technical Manual, TM5 632, October 1945.)

whole area of pest control) taught us that water management is the key to mosquito suppression. This is as true today as it was then.

"The most important aspect of mosquito control is source reduction," says Minoo Madon of the California Department of Health Services, "which means preventing the mosquito from emerging as a mosquito, before it becomes a nuisance."

While public health authorities have the responsibility to prevent mosquito breeding on public lands, you as a householder can do a great deal to help.

Metropolitan Areas: Private Breeding Sites

If you live in an urban area, the mosquitoes feeding on you are most likely breeding in artificial containers or other small sources right near your home, some even on your own property. To launch your personal mosquito control program, make a stagnant-water survey around your home and eliminate anything or any place that collects water. Correcting all of these is usually a simple matter.

Empty jars and tin cans should be put out for trash collection. Old tires should be discarded. If your community does not pick up discarded tires, or the landfill refuses them, keep them covered and dry. If you don't have a shelter, use a tarpaulin. Worn-out tires are major mosquito breeders across the country. Keep tire ruts filled with dry dirt or pave your driveway.

Get rid of all abandoned toys—wagons, trucks, doll carriages, sand pails. Your town may have a toy loan library that will refurbish these and lend them to youngsters. Empty your children's wading pool at least every 3 days in summer, and keep it in a dry place the rest of the year. Just turning it over won't do much good since water accumulates in the sag during a rain.

Screen any cisterns or rain barrels.

Empty the bird bath every 3 days.

Change your pets' drinking water daily.

Store wheelbarrows on end and turn empty plant tubs upside down.

If the tubs have drainage holes in the bottom, leave them on their sides so that water doesn't seep up through the holes.

When using plastic sheeting to control weeds, be sure to drain it at least every 3 days.

Root your outside plants in sand, dirt, or vermiculite. Many people root plant cuttings in water; that's fine indoors, but outside the water and vegetation are what an egg-carrying mosquito looks for.

Drain water from tree holes. Fill the holes with sand or cavity cement.

Or, drill a small channel from the bark to the bottom of the hole so that water drains away.

Monitor pools on your property. Although a delight to the eye, an ornamental pool can be a mosquito breeder for the whole neighborhood. Don't despair. You can still enjoy your water lilies. Just check with your local mosquito abatement district to see if they supply mosquito fish to private parties. The fish, a kind of minnow, has a voracious appetite for mosquito larvae. Another possible source for the fish may be a nearby arboretum or your state's fish and wildlife department.

A properly filtered and cleaned swimming pool will not encourage mosquitoes, but a poorly drained pool deck will. If water regularly collects around your pool, you may need to have the grading and paving corrected.

Drain and overturn any small boats on your property.

Roofs and Gutters

Your house may have some construction faults that are attracting mosquitoes. Sagging roof gutters or a flat, undrained roof can bring insects needing a lot of moisture. A sheetmetal contractor can fix your gutters and downspouts. Although it's a small job, you may need a general contractor to bore a hole in your flat roof for better drainage.

Other Checkpoints

Three other possible mosquito breeders around a house are leaky outside faucets, puddles that gather under air-conditioners or other house coolers, and the vault of your water meter. The leaky faucet may only need a new washer. Set a container under your cooler and empty it regularly. Put a couple of inches of gravel into the water meter to make it useless to a pregnant mosquito. It will also discourage cockroaches.

Metropolitan Areas: The Streets Above and Below

Nearby public or industrial facilities may be breeding more mosquitoes than any water you and your neighbors are neglecting.

Storm Drains

Scientists of the University of California at Riverside suspect that most of the mosquitoes in Los Angeles County are "undergrounders," bred in the thousands of miles of storm drains and catch basins criss-crossing southern California. This is probably true in most urban areas. These researchers' plan to wipe out the Los Angeles mosquitoes underground includes introducing guppies and mosquito fish to consume the larvae.

Gutters and Potholes

Street gutters that retain water runoff are another mosquito source, as are polluted ground pools, cesspools, open septic tanks, and effluent drains from sewage disposal. Chewed-up roadways, pocked with potholes and cracks, can nurture hordes of mosquitoes. A pint of floodwater can support 500 lively wigglers, and many a pothole holds considerably more than a pint. If roadways and street gutters in your part of town need repair, and city leaders are sitting on their hands, round up your neighbors to join you in a well-publicized complaint.

Cemeteries

Honoring the dead can be dangerous to the living. In Caracas, Venezuela, a recent epidemic of dengue fever was traced to the city's cemetery. There, grieving relatives keep fresh flowers in water-filled vases at the graves. The water is replenished when the flowers are changed and when it rains. More than 190,000 vases are launching 50 million mosquitoes over the city. Although health workers have recommended substituting artificial flowers for fresh or filling the containers with soil, the ways of death do not easily change. Caracas's health menace remains.

If you live near a cemetery, try to have its managers encourage plastic flowers, potted plants, or small flags as grave decorations.

Technology's Throwaways

Tires. The industrial world's advanced technology and quick replacement of worn-out parts are hindering the ongoing war against mosquitoes. The most serious problem is worn-out tires. One water-filled cast-off in La Crosse, Wisconsin, for example, was found nurturing 5000 mosquito larvae, a dangerous number carrying encephalitis. Some 240 million tires are discarded in the United States every year. Because tires tend to rise to the top of a sanitary landfill, many city dumps don't accept them, so they're just tossed onto vacant land. There they collect water, ready for any egg-carrying, disease-carrying mosquito.

All-rubber tires are easily recycled, but steel-belted radials resist recycling in a cost-effective form. Environmental specialists are working on this problem and turning up some ingenious solutions. For instance, chained together, hundreds of tires form an effective barrier reef for an oceanside community. Or, weighted, chained, and dropped into a freshwater lake, the tires make a fine spawning ground for catfish and bass.

If your community has no satisfactory way of disposing of old tires, store them covered, either in a dry shed or under a tarpaulin. (For using them safely for children's swings, see the section on spiders in Chapter 13.)

Abandoned Automobiles, Refrigerators, and Buildings. Derelict cars and discarded refrigerators, especially those with plenty of organic matter inside, can produce swarms of mosquitoes. Deserted buildings are another fertile source. Containers gathering water outside abandoned structures, dilapidated roofs and gutters will all attract the pest.

Industrial Plants. Living within a mile or two of a factory can aggravate your mosquito problem. Industries often use great volumes of water and also maintain sumps, pits, and water towers. Improperly drained industrial waste water can launch squadrons of mosquitoes to nearby homes.

Whether the nuisance is a house, a store, a refrigerator, a car, or a carelessly run factory, protect your family and yourself with a call to public health authorities.

Rural Areas

As subdivisions spring up farther and farther out from the cities, the mosquito becomes more of a pest. Suppressing it in a rural or semirural environment is more complicated for the individual resident than it is in an urban setting. Besides having to watch all the trappings of family living, farmers and rural householders should be aware of the many agricultural processes that foster mosquitoes.

Dairies

A dairy uses huge amounts of water to wash its cows, barns, and equipment, water rich with organic matter that nourishes mosquito larvae. Often it's dumped on impervious ground, where it forms stagnant pools. If a nearby dairy is sending clouds of mosquitoes or even just a few aggressive females your way, let your local mosquito abatement district know about it. They can have the situation corrected.

Irrigation Facilities

As we've moved to drier, warmer climates, we've developed vast irrigation networks to turn these former deserts into bountiful farms. Where water doesn't drain properly, standing in ditches, laterals, and canals, or gathering in pools below the watered fields, ten irrigations a year can produce ten generations of mosquitoes in the same 12 months. "Swarms can become so dense, that one's clothing becomes covered with a continuous layer of insects," says entomologist Walter Ebeling of the University of California, "and livestock may be killed by their continuous attacks."

Ponds, Watering Holes, Hoofprints

There can be other mosquito sources in rural areas, such as hoofprints around watering troughs and livestock watering holes with plants clustered around their banks. Troughs should be set on gravel or cement to prevent pooling. Any ponds should be deep and straight-sided to discourage plants where mosquito larvae can hide from predators. A population of greedy mosquito fish will also keep the pond clear of wigglers.

If you run a farm or dairy, or live near one, that is having a mosquito problem, your county agricultural agent or mosquito abatement district can help you eliminate the pest.

Getting a Mosquito Abatement District Started

You've gathered by now that mosquitoes are very much a public health concern and controlling them a public responsibility. Many towns and cities have established agencies to keep the insect within tolerable bounds. If you live in a small community that has no central mosquito control agency, your best long-range solution is to start a mosquito abatement district. *Every one of the fifty states has legislation permitting localities to establish abatement districts.*

To form such a body, citizens petition their city council or county commissioners to set up a special tax for funding a mosquito control agency. The bureau is staffed by a manager, a source-reduction expert (someone who can spot a mosquito breeding site and knows how to dry it up), an entomologist (insect expert), and inspectors and supervisors to see that the necessary drainage or chemical applications are carried out.

Allies in Mosquito Control

Forming such a district takes time but after you've submitted your petition, you don't have to sit idle and just endure mosquitoes. Many organizations have a real interest in maintaining your community as a comfortable, healthful place in which to live and do business. As a home-owner and taxpayer, you can call on the local chamber of commerce, the realty board, women's civic groups, city and county health departments, schools, the planning commission, and the street and road department.

If you live in a rural area, farm bureaus, dairy leagues, cattleherders' associations, and the grange all have a vital stake in suppressing mos-quitoes. Urge their members to help you publicize the problem—and its solution—through newspaper stories, radio and television programs, and school classes or meetings. Figure 7.2 is a leaflet used by the national Centers for Disease Control alerting communities to the need for mos-quito control and telling them how to do it. You may copy it and use it.

Mosquito-Disease Emergencies

What if encephalitis or dengue fever or some other mosquito-borne disease breaks out in your area and there's no public agency responsible for mosquito control? What can you as a citizen do?

You turn to your state. It is empowered to protect its people's health. Request that entomologists and other public health specialists be sent to your community to monitor and suppress the outbreak. If necessary, your state officials can call on the Centers for Disease Control in Atlanta, who will send in experts to launch the needed program.

How to Protect Yourself

All community control measures boil down to one question: How can individuals defend themselves from the mosquito out for blood? Despite massive government efforts at controlling the pest, it is always with us. As we've seen, the chemical pesticides are weakening in their ability to protect us. What can a person do?

HELP STOP MOSQUITO BREEDING

Dump Standing Water to Kill Eggs, Larvae (Wigglers), and Pupae (Tumblers).

Eliminate All Standing Water in Which They Can Breed.

Wash and Refill Animal Drinking Pans Daily

Grow House Plants in Earth, Sand, or Vermiculite . . . Not Water.

CORRECTIONS NEEDED

——None. You have the thanks of your community and neighbors.

——Clean up and dispose of cans, bottles, old tires, or other containers which may hold water.

For additional information call ———————————————

U. S. DEPARTMENT OF HEALTH AND HUMAN SERVICES
Centers for Disease Control
Atlanta, Georgia 30333

Figure 7.2. (*"Help Stop Mosquito Breeding."* United States Department of Health and Human Services. Atlanta: Centers for Disease Control, Leaflet.)

Source

empty cans, jars	____	wheelbarrows	____
watering cans	____	plants rooting in water	____
old tires	____		
tire ruts	____	plastic sheets to control weeds	____
abandoned toys (trucks, wagons, doll carriages)	____	ornamental ponds	____
wading pools	____	tree holes	____
bird baths	____	sagging roof gutters	____
cisterns	____	flat, undrained roof	____
rain barrels	____		
pet water dishes	____	swim pool deck	____
puddles under house coolers	____	leaky outside faucets	____
small, undrained boats	____	water meter vault	____

Shut Them Out

At Home

First, use the preceding checklist to eliminate any possible mosquito nurseries on your property. Check off each stagnant-water source that you clear up.

Screens

Examine all screens carefully. They're your chief line of defense against flying pests. Yours should be in good repair, with no tears in the mesh, and should fit snugly in the door or window frame. Since some species, like the western treehole mosquito (a disease carrier), are only $\frac{1}{16}$-inch long, you may need a finer-gauge mesh than the normal 16 by 16 by 20. A fabric 16 by 20 by 23 keeps out the smaller breeds.

Screen doors should open outward to prevent any mosquito from pursuing you into the house, and they should close completely within 4 or 5 seconds. Make certain too that all windows and outside doors fit tightly.

And Abroad

In many countries, especially the less developed ones in warmer climates, screens are a luxury beyond the means of most people. When visiting these places, be aware of the lack of screens and the resultant much greater risk there of mosquito-borne disease. Dengue fever, for example, occurs fairly frequently in the West Indies and the South Pacific.

Before going to a warm country, check with your doctor or public health department about immunization against illnesses that could ruin your holiday. For some mosquito-transmitted diseases there are no vaccines and no specific medicines. Your best, indeed your only, protection is a good repellent.

Chemical Repellents

Deet

The most effective mosquito repellent, sold as a spray, stick, or lotion, is Deet, the popular name of diethyl toluamide. In commercial products it's often mixed with substances repellent to insects other than mosquitoes, thus giving good all-around protection.

Deet was used by American troops in Vietnam, and is the standard against which the United States Department of Agriculture measures all newly developed repellents. According to the Montana Department of Health, products containing a higher percentage of Deet do a better job than those with less of the chemical. Deet will keep mosquitoes at bay for anywhere from 2 to 12 hours. It's sold commercially as Deet or Off and is included in other products as well. Two other good chemical repellents are dimethyl phthalate and Rutgers 612.

Proper Use of a Chemical Repellent

Deet should be applied directly to the skin. It will not damage nylon so it can be applied over hosiery. You can also spray it on clothing to repel chiggers, fleas, ticks, and midges. However, it can damage plastics such as watch crystals, pens, and rayon fabrics. It dissolves paints, varnishes, and leather. On a recent trip through the mid-Atlantic states I learned not to carry Deet unprotected in a leather handbag. A leaky applicator ruined my purse.

When applying a repellant—any repellent—on a child, don't put it on the youngster's hands. These substances irritate sensitive tissues, and children often pop their hands in their mouths or rub them over their eyes.

Natural Repellents

Before the chemicals appeared, our grandparents and their parents used various natural substances with good effect to keep mosquitoes from biting.

Oil of Citronella

This derivative of a lemon-scented grass grows in Java and Sri Lanka. It has long been a reliable barrier against insects. Burned in candles out of doors, its fragrance lends a pleasant spiciness to the evening air. Many humans like the aroma, but mosquitoes don't.

Garlic

Garlic, that bud beloved of country people for any ailment from the common cold to nervous disorders, is recommended by Edward E. Davis of the Stanford Research Institute as a mosquito repellent. He suggests eating a lot of garlicky food. Of course, in addition to holding off mosquitoes, it may also hold off some of your friends. You have to decide if the trade-off is worth it.

Garlic may also be an answer to our larger mosquito problems. In recent experiments, a highly concentrated extract of it killed all the mosquito larvae exposed to it.

Pennyroyal Mint, Tansy

Other natural repellents laypeople have found helpful are oil of pennyroyal mint rubbed on the skin and tansy planted near a door. Oil of pennyroyal, however, can cause a rash. Incidentally, if someone tells you to plant a castor bean plant near your door to repel mosquitoes, ignore the advice. The castor bean contains ricin, a poison used in recent political assassinations.

Basil

Several garden books describe basil as a mosquito repellent. I can testify that it is. For years I was imprisoned indoors on summer evenings by my attractiveness to mosquitoes. Now, with basil plants flourishing along the edge of our patio, I'm able to spend a serene evening outside, free of the insect's attacks.

Electronic "Controllers"

Various devices on the market claim to control mosquitoes. They range from the pretty useless to the utterly useless.

Electronic Repeller

When turned on, this gadget gives off a high-pitched sound. In scientific tests it proved worthless. Whether the "repeller" was turned on or off, the mosquitoes bit whoever was holding it.

Insect Electrocutors

Called Bug Blaster, Bug Whacker, Zapper, and the like, insect electrocutors do kill flying insects. They attract the creatures with ultraviolet light and then electrocute them. These products, however, are not much help against mosquitoes.

In tests conducted in 1982 by scientists from the University of Notre Dame, only 3.3 percent of the insects killed by these devices on an average night were female mosquitoes. Human beings near the electric insecticides proved to be more attractive than the ultraviolet light, and people had just as many mosquito bites as before the appliances were installed. The device, which in 1984 cost anywhere from $40 to $130, also kills valuable predators, and may actually attract more insects to your yard than you would otherwise have.

Nature's Controllers

Nature has some potent mosquito controllers that have been on the job for millions of years.

Birds, Reptiles, Predatory Insects

Purple martins, whippoorwills, and hawks all eat mosquitoes. In addition, mites, ants, and booklice relish mosquito eggs, while frogs, turtles, and a large, harmless species of mosquito prey on the pest's larvae and pupae. Spiders, mites, geckoes, lizards, and bats devour adult mosquitoes.

Dragonflies make short work of both mosquito larvae and adults. A number of towns in Maine are buying dragonflies by the thousands, following the lead of the town of Wells, which has used these insects for years as a major means of mosquito control.

The problem with all these predators is that they may not eat the mosquitoes we humans want eaten. We can't confine the birds, bats, dragonflies, and lizards to the areas we think need cleaning up.

Mosquito Fish

Gambusia minnows, or mosquito fish, are used extensively in ponds to keep down mosquito larvae. It is the one predator that we can confine exactly where it will do the most good. Many mosquito abatement districts supply them to citizens free of charge. The city of Colorado Springs, Colorado, bases much of its mosquito control program on this minnow.

Possible Future Controls

Scientists have been investigating other natural controls for this eternal nuisance, controls that pose no danger to other life forms. One of the most promising is a bacillus developed in Israel that infects mosquito young with a fatal disease. Another substance that blocks the maturing of larvae is also available. Both of these, however, are best suited to wide applications like large accumulations of water. Still a third technique under study is the release of sterile males.

Certain algae and freshwater fungi have been found to be poisonous to immature mosquitoes. The gluey seeds of a common California weed are being tested for their ability, when placed in water, to attract mosquito wigglers, which then stick to the seed and die from drowning or exhaustion. One pound of the seeds is estimated to wipe out 54 million larvae.

Why the Mosquito Bites People

Why is it that in a circle of friends chatting outside on an evening, one or two may be heavily bitten while the rest sit comfortably, apparently immune? Scientists tell us that there are malaria and yellow-fever carriers who would rather feed on a chicken, a cow, a pig, or a horse than a human but, lacking the livestock, will settle for a man or woman.

As far as researchers can determine, the combination of warmth and moisture are very alluring to mosquitoes. So is carbon dioxide. Dry ice, a form of the gas, is used to trap the insect when scientists want to know what species are in an area. Since we give off carbon dioxide with every breath, all of us are potential mosquito meat.

However, other factors enter the picture. Some species like light-colored clothing, some dark. The yellow-fever–dengue mosquito prefers a man to a woman, and an ovulating woman is more likely to be bitten than one who's menstruating. This seems to be tied to the amount of estrogen in the blood. All in all, warmth and carbon dioxide, enhanced by amino acids and estrogens, appear to determine our appeal for mosquitoes.

If you're someone mosquitoes find delicious, here's how to avoid being bitten:

1. Stay indoors when mosquitoes are active. This may be day or night, depending on the species.
2. Use a proven repellent (see above).
3. When outside, wear long-sleeved clothing made of a tightly woven fabric. Keep sleeves and collars buttoned, and trousers tucked into socks. This is the method used by public health workers in areas known to harbor disease-bearing insects.

Water Management Still the Sure Preventive

Water management; monitoring the number and kinds of mosquitoes in an area; strategic use of predator fish, screens, and repellents; and expert use of selective pesticides by trained professionals are all used today with good effect.

But despite all our science and technology, the insect remains a major threat to human health and comfort. Mosquitoes abound in such astronomical numbers that we will probably always be warring with them. To keep the fight even, we must be ever alert to the ways we handle water, from the smallest outside puddles to the vast salt marshes rimming our coastlines. Strict water management will enable us to coexist safely with this graceful, persistent, potentially deadly enemy.

CHAPTER 8

The Mighty Flea, the Insidious Tick

Fleas

The seventeenth-century Dutch artist who sketched the first flea seen under a microscope was heard to mutter while peering through the lens, "Dear God, what wonders there are in so small a creature!" He spoke the truth.

Looking rather like a flattened armor-plated tank, the wingless insect has strong "thorns" and hooks on its head and thorax that make it hard for its furry host to pull it off. Its powerful legs can shoot it 7 or 8 inches up into the air, 14 inches across the level. A human with such strength could clear the top of the Washington Monument or Egypt's Great Pyramid at one bound.

The flea can pull 400 times its own weight and lift an object 150 times heavier than itself. By contrast, a horse can pull 4 times its weight. In addition to its great strength, one authority claims the insect can be frozen for a considerable time and thaw out good as new.

Are Fleas Dangerous?

Although it has fallen behind the mosquito as the world's most dangerous insect (mosquito-caused deaths total millions annually, those from

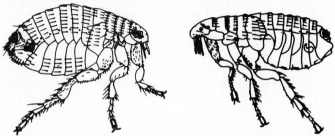

Figure 8.1. Two types of fleas. (*Fleas of Public Health Importance and Their Control*. Atlanta: Centers for Disease Control, Homestudy Course 3013-G, Manual 7-A, 1982.)

fleas, thousands), the flea still is a catastrophe waiting to happen. Throughout history the plague that it harbors has periodically depopulated whole nations, even continents. In the sixth century A.D., plague killed almost half the people in the Roman empire. Eight hundred years later, probably helped along by the Crusaders traveling between Europe and the Middle East, an estimated three-fourths of the people in Europe and Asia succumbed to it.

Although plague attacks and kills many fewer people today, it is still a danger, more so in the third world than in industrialized nations. However, the United States west of the 100th meridian is one of the world's major reservoirs of this fearsome disease. In just 2 months—from April to June 1983—there were sixteen cases in Arizona, New Mexico, Utah, and Oregon. Plague-infected squirrels have been found within Los Angeles' city limits, and there have been human cases in nearby suburbs.

The infection can take one of three forms: bubonic, pneumonic, septicemic. Since the development of antibiotics, bubonic plague, the dreaded Black Death, is much less deadly than formerly. Untreated, however, it has a death rate of 25 to 50 percent. Pneumonic and septicemic plague are almost always fatal.

Once a scourge encountered in seaports with their ship-infesting rats, plague is now primarily rural, carried by the parasites of chipmunks, squirrels, opossums, and other small mammals. A cat living in or near a wilderness area can readily pick it up from these animals and transmit it to humans.

Such a case happened in Lake Tahoe, California, in 1983, when a young schoolteacher contracted pneumonic plague from her cat and was diagnosed too late for medicines to help her. Fortunately, authorities were able to protect her pupils with antibiotics.

If you live in or near a western wilderness area and have pets, keep them confined and rodent-proof your house as described in Chapter 5. And, of course, don't make friends with any ground squirrels and chipmunks.

Other serious diseases fleas carry are typhus, tularemia, and tapeworm; the insects can also provoke strong allergic reactions.

Controlling Household Fleas

Controlling disease-bearing fleas and their rodent hosts is a problem for public health authorities. Controlling fleas on your pet and in your home is your problem, one that can be a real headache. They're among the most stubborn nuisances we encounter.

Thousands of people make a good living turning out antiflea powders, lotions, shampoos, and collars; dipping infested pets; compounding medicines to heal animals who've ripped their itching skin in frenzy; spraying our homes and yards with chemicals to stem the flea onslaught. At best, they've had only modest success. Even fumigating a house gives only short-term relief.

What makes this pest so invulnerable? The secret lies in the insect's life cycle, its ability to survive for long periods without food, and its readiness to move from one kind of host to another.

Nature of Fleas

Species of Fleas

Of the estimated 1600 known kinds of fleas, only 3 are important for the average person: the dog flea, the cat flea, and the human flea. In addition to these 3, the rat flea, chief carrier of plague, is found wherever there are rodents. Lately, with a rise in the number of plague cases, we've had to pay more attention to the rat flea. Although the parasite prefers to feed on the animal from which it takes its name, each of these species very quickly shifts to another mammal when its host dies.

Cat and Dog Fleas

Cat and dog fleas infest many other animals besides our pets, including chickens, rats, opossums, raccoons, and foxes. They readily bite people and are the ones most likely to bother you. Their larvae prefer an environment of dust and organic debris and are very common in yards and under houses.

The Human Flea

At one time the human flea was the type most commonly annoying people. An upper-class woman of the Middle Ages often passed the time searching herself for fleas, and would bestow on a favored suitor the privilege of helping in the hunt. Because of its exceptionally strong legs— even for a flea—this was the star of the flea circus, glued for life to a tiny chair or tethered to a lilliputian coach or cannon.

Formerly homemakers damp-wiped their wooden floors and furniture, providing the flea with the humidity essential for its reproduction. With the advent of the vacuum sweeper, this flea is now found mostly on pigs. Its bite can cause a severe allergic reaction.

The Chigoe ("Chigger")

If you live in the tropics, you may have the bad luck to meet up with a chigoe, a nasty flea that burrows between people's toes and under their toenails, causing intense pain and, ultimately, even gangrene. It's the inspiration for the expression, "I'll be jiggered!" The best cure for chigoe bite is not to walk barefoot.

"Sand Fleas"

So-called sand fleas are actually several species. In the northern United States, they are cat or dog fleas found in vacant lots, where they've dropped off stray animals. In the west, they're cat or human fleas that have fallen from deer, ground squirrels, or prairie dogs. In the south, they may be poultry fleas (also called "sticktights") or cat or dog fleas. On the beaches, sand fleas are harmless crustacea rather like shrimp.

Snow Fleas

Snow fleas are not fleas at all. They're springtails. A springtail is a tiny, soft-bodied, wingless insect. By suddenly releasing an appendage tucked under its abdomen, it can spring into the air, causing it to be mistaken for a flea.

The Flea's Life Cycle

The female flea lays her eggs either on or off the host animal. Those laid on the host roll off to the animal's bedding or nest, or accumulate in carpets, floor cracks, furniture, dust and—outside—in damp soil, manure, and other organic debris.

Depending on temperature and humidity, the larvae hatch in anywhere from 2 to 14 days. They prefer temperatures between 66 and 84°F and humidity between 70 and 90 percent. Flea eggs never hatch when the thermometer falls below 40°F and most larvae die when the humidity is less than 40 percent. This is why a wet winter or spring usually foreshadows a bad flea season.

"In warm weather the average dog has sixty fleas," says Dr. Steve Wagner, president of the California Veterinary Association, "half of them female. A flea can lay 600 eggs a month. That's 18,000 eggs a month."

The larvae feed on organic debris in the cracks where the eggs usually hatch, in carpet dirt, under beds, and in basements. The young of some species feed on the dried blood excreted by the adults. The "pepper and salt" often found where an infested animal sleeps is a mix of flea eggs and this dried blood.

A Long Infancy

Anywhere from 9 to 200 days after hatching, the larva spins a cocoon and becomes a pupa. This stage can last from 1 week to 1 year, with the adult flea often resting inside the cocoon for many months, ready to pop out when it senses vibrations, warmth, or carbon dioxide from a nearing host. The ability to prolong its life span is at the root of the flea explosion in summer, when warmth and humidity are just right.

It's also the reason why many people with cats or dogs, on returning from a vacation during which their pet was boarded, are attacked by hordes of hungry fleas. And the fleas don't care who owns the pets.

A Florida man was plant-sitting for pet-owning neighbors on a holiday and, while watering the houseplants, felt a stinging sensation over his legs. Looking down he saw that his calves and ankles were covered with feeding fleas. It's a universal experience. In Arab countries it once was the custom to pay an unemployed man to enter every room of an uninhabited house to lure all the fleas from their hiding places. Canny country folk in the southern United States will set a sheep in an empty building as flea bait.

Shut Fleas Out

Preventing Fleas

Preventing fleas is probably easier than getting rid of them. The simplest way, of course, is not to have a pet. Those for whom a loyal dog or affectionate cat is one of life's great pleasures almost inevitably will face a flea problem at some time in their animal's life unless they establish and hold fast to some ground rules when they first bring home their puppy or kitten. Here they are:

1. Keep your dog outside at all times. Provide it with a sturdy dog house and a well-fenced *concrete* exercise yard.
2. Don't let your dog roam the neighborhood. It can pick up fleas from infested gardens or other animals.
3. Decide at the outset if your cat will live inside or outside. Most people whose cats come in and out at will have flea problems.
4. In warm weather comb your pet every day with a flea comb.

Flea Sources Other than Pets

Even if you don't keep a cat or a dog, you may still find yourself coping with fleas. Where are they coming from? There are several possibilities.

1. If you've just moved into the house, the previous occupants may have had a pet.
2. An infested cat may have had her litter in the crawl space under your house or porch. (When the kittens went off on their own, the flea eggs hatched and the adults are looking for the nearest blood meal.)
3. Maybe a squirrel or other rodent once nested in your attic and when it was excluded, its fleas and their young remained behind.
4. In rural areas, the source could be an opossum in the wall voids of your home.
5. In city or country, infested rats in the wall voids or nearby could be the origin.

Eliminating these would be an effective first step in getting the situation under control.

Wipe Fleas Out

Insecticides Can Be Cruel Protection for Pets and Owners

The cat or dog being sprayed or powdered with chemical insecticides is at considerable risk of poisoning. The animal's owner may be too. Cats and dogs, says Dr. William B. Buck, veterinary toxicologist of the University of Illinois, are frequently poisoned by organic insecticides. The chemicals most commonly used in antiflea collars, dips, sprays, shampoos, and powders contain organophosphates and carbamates. These substances cause most of the pet poisonings.

Occasionally a pet owner will be so intent on suppressing fleas, his or her animal wears an insecticidal collar and is sprayed and powdered regularly. All three can be too much and the veterinarian ends up seeing a seriously ill patient. Or the collar-wearing dog or cat may need worming or develop mange; treatments for these include pesticides and the resulting dosages may be heavier than the animal can take.

Antiflea sprays are compounded with a solvent, itself often toxic, that enables the insecticide to penetrate your pet's skin—and yours when you stroke your pet. Powders and dusts have no solvents, but you and your pet breathe in the insecticides when they're being applied; and afterward the poisons can drop onto floors, carpets, and furniture. A cat, frequently licking its fur, also ingests a good deal of the powder, and in its comings and goings may pick up insecticides from neighbors' yards, greatly increasing its toxic intake.

Flea bombs that are so popular and seem to offer such a quick solution, can make humans ill if improperly used.

Other substances may be highly questionable. Internally given pesticides kill fleas that bite the animal. However, they could also kill your pet. These chemicals were developed for 1000-pound cattle and may not translate safely to a 10-pound cat or a 40-pound dog. Their potential for seriously harming a dog is great (*they should never be given to cats*). Their long-term effects are unknown and they may contribute to kidney and liver disease.

There are also substances that block the development of flea young.

As with many insecticides, scientists are uncertain about their long-term effects.

Flea collars are frequently put on new pets as soon as they are brought into a home, so anxious are the new owners not to have fleas. What people are doing is giving their darling a dose of toxic vapors 24 hours a day. And every time they play with the dog or cat, they're getting some of the fumes too.

The collars may be impregnated with methyl carbamate or an organophosphate, poisons that can be absorbed directly into the blood stream. Many animals, especially cats, develop dermatitis on their necks from these irritating substances. The dermatitis can last for months, says entomologist Walter Ebeling. Humans can also develop rashes from them.

The collars, according to veterinarian Wendell O. Belfield, are worthless. "The flea collar may kill a couple of insects a day," he says, "but others by the numbers are thriving nearby in the carpet or yard." Nor do the collars work well on large dogs since the vapors evidently don't reach the tail, where fleas tend to gather. An herbal collar, safer than an insecticidal one, may be no more effective.

If you feel you must use a chemical, the safest, according to the Centers for Disease Control, are pyrethrin dusts. But these may only temporarily paralyze the insect. A good way to go with these botanicals is to stand your pet on newspaper while powdering it, comb the stunned insects onto the paper, and then promptly burn it. You'll need about 1 ounce of dust for a large dog. For adult animals use a 1-percent dust; for kittens and puppies, 0.2-percent. Pyrethrins may stimulate the fleas for a short time, making your animal temporarily uncomfortable.

A final word on flea insecticides. They may not bother the fleas at all, but you may be allergic to them.

A few years ago a delightful cat adopted us. She got all of our attention, that is, when we weren't scratching flea bites. Although I powdered her every week with a carbamate the veterinarian had sold me, the fleas just grew fatter and sassier.

Unwilling to fill the house with the fumes of a flea bomb (a friend had been made sick by one), I bought a highly recommended spray and treated cracks, rugs, and other flea hideouts. I applied the chemical strictly according to the label's directions. It did the trick. The fleas were completely routed, but our son nearly was too. He had a strong allergic reaction to the product, with swollen eyes, puffy lips, and difficulty in breathing.

Alternative Flea Control

Considering the flea's incredible jumping ability, its widespread and varied hiding places, and our animals' fondness for roaming, it's probably impossible to have a pet that's always free of fleas. However, to keep your pet and you comfortable without exposure to pesticides you have three overall strategies:

1. Environmental controls (both indoors and out)
2. Repellents
3. Mechanical controls (such as traps and bathing)

No single one will bring you complete success and, because animals and people vary greatly in their sensitivity to insect bites, you'll have to experiment to see which combination works best for you. While searching for that mix, don't get discouraged. At least 40 percent of the commercial products sold over the counter for flea control are ineffective, according to veterinarian Steve Wagner.

Environmental Controls

Outdoors

"If you're talking flea control," says veterinarian George Peavey, "you're talking environmental control." Fleas can be in your upholstery, carpets, the cracks and crevices of your parquet floor, and in your yard. They may also be in your attic, basement, and wall voids. (Remember, even if you don't have any pets, you could have a flea infestation.)

- Check the immediate area for rodent nests and destroy any that you find. They may be empty but fleas can still be hiding in them.
- Since fleas rest in organic material, use overwatering or drying to kill them in your yard, advises the Center for the Integration of Applied Science in Berkeley. The Centers for Disease Control suggests that you remove any animal manure and organic debris promptly.
- Make sure that the outside structure of your house is in good repair. Foundation, foundation sills, windows, doors, and vents should all be tightly sealed. Inspect your chimney and the crawl space under the subflooring, and close off any cracks and holes.
- Rodent-proof your home as recommended in Chapter 5. Be espe-

cially careful if you have an attached garage, a frequent entry for flea-carrying rats and mice. Make sure no stray cat can find a niche for giving birth.

- If the fleas are coming from a stray animal, you may need to treat your attic and wall voids with a long-lasting silica gel. This material is very irritating to the lungs so when applying it wear adequate protective gear.

- In summer, when fleas are at their peak, you may have to keep your pet outside all the time. Keep the animal comfortable by regularly vacuuming its sleeping place and laundering its bedding frequently.

Indoors

- Vacuum any infested rooms of your house frequently, daily if necessary, using the crevice attachment to suck up insect adults, larvae, and eggs from upholstery crevices and carpet corners, loosened baseboards, and floor cracks. Place each day's collection of dust and insects in a tightly closed bag and either burn it or set it to "bake" in a covered garbage can in the hot sun.

- To bring a heavy infestation down to tolerable levels quickly, have your carpets and upholstery steam cleaned.

- If your pet sleeps inside, cover its resting places with removable, washable cloths. Wash them every couple of days since flea eggs can hatch 2 days after being laid.

In the 1930s, Hugo Hartnack, an early practitioner of integrated pest management had these suggestions for fighting fleas. Few of us sleep on straw mattresses, but his other ideas are still good.

Have a cement-floored basement.

Don't use straw mattresses.

Have crack-resistant, smooth hardwood floors.

Clean your uncarpeted floors with oil instead of water.

Vacuum all rug and carpets frequently.

Use central heating to dry the air in your home.

Don't keep pets; get rid of strays.

Note: His final suggestion would be useful before you change homes. Heat your new residence to 122°F (for several hours) before your furniture arrives.

Repellents

Over the years people have found several of the following flea repellents effective, at least to some degree. Veterinarians may tell you they don't work, but they see animals with serious flea allergies. A cat or dog without fleas, or untroubled by their bites, doesn't need a veterinarian for flea allergy.

Brewer's Yeast

Veterinarians are divided on the value of brewer's yeast. Some say it's useless, others think it's well worth a try.

Dr. Wendell Belfield, while not claiming miracles for brewer's yeast (if you've read the rest of this book, you know there are no miracles in pest control), says some cats and dogs respond well to this rich source of the B vitamins. He suggests starting to give it in spring and continuing all through the warm weather to build up and maintain your pet's flea resistance. The yeast apparently imparts an odor to the animal's skin that fleas don't like. Dr. Belfield suggests a daily dose of 25 milligrams per 10 pounds of the animal's body weight.

A Word of Caution: Brewer's yeast given in large doses or with dry food swells, causing the animal considerable intestinal discomfort. We mixed $\frac{1}{2}$ teaspoonful daily in our cat's moist food for months. She accepted it readily and had no cramps or gas.

Some cats and dogs are allergic to yeast. One vitamin manufacturer, Schiff, has a line of rice-based B-complex vitamins that do not bother yeast-sensitive animals. These products are available in health food stores.

Herbal Repellents

There seem to be as many herbs suggested for flea control as there are pet owners. Everyone has a favorite. Some herbs are mentioned more frequently so these ought to be the first you try. Many people claim that pennyroyal mint, eucalyptus, citronella, and rosemary sewn into a collar worn by the animal will hold off the pests. The oil of the Australian tea tree, which is notably free of insect pests, is another one mentioned, as is dried wormwood. According to Dr. C. E. Schreck of the United States

Department of Agriculture's Insect Repellent and Attractant Project, southerners plant wax myrtle near their homes' foundations to repel the fleas. Dr. Belfield recommends rubbing ground cloves or eucalyptus oil into your pet's fur as an alternative to an insecticidal collar.

Lemons

An infusion of lemons may work for you. In San Jose, California, Mrs. Dee White, has kept her three dogs free of fleas for years using it regularly. Here's her recipe:

Cut four lemons in eighths. Cover them with water, and bring to a boil. Simmer for 45 minutes. Cool and strain the liquid. Store it in a glass container (it may seep through plastic).

Wet your animal thoroughly with the infusion, brushing its coat while wet so that the lemons' juice and oil penetrate down to the skin. Dry the fur thoroughly with towels and brush again.

There is sound basis to this unusual approach. University of Georgia scientists have found that grated citrus peel kills a number of household pests, including fleas.

Commercial Repellents

Flea Ban II is a commercial product that seems to work. We kept our cat almost completely flealess during one of the worst flea seasons in years by using only daily grooming with a flea comb, brewer's yeast, and Flea Ban II. (Flea Ban I is a shampoo.) So mild that its developer, chemist Terry Jones, takes a swig of it during demonstrations, Flea Ban II can also be used safely on puppies and kittens. Saturate the animal's fur for greatest effectiveness. You can also spray it on upholstered furniture and drapes after testing for colorfastness. It's sold with a money-back guarantee and distributed by J & J Products, 5144 Crenshaw Boulevard, Los Angeles, CA 90043. A final note: Terry Jones says Flea Ban II tastes terrible.

Traps and Other Mechanical Controls

People have been applying their ingenuity for a long time to the problem of flea control. Someone in the eighteenth century even devised a trap for human fleas, to be worn around the neck. It probably didn't work.

A more recent trap needs an accommodating woman. Based on the sex's reputed attractiveness for fleas, a woman wearing slacks smeared with a sticky substance walks through a flea-infested room. The insects, aroused by her warmth and body vapors, will jump on her slacks and be stuck.

The Flea Comb

More reliable than either of these is the flea comb, available in pet stores. When combing your pet, have a container of soapy water nearby for drowning your catch. Be sure to get the comb's teeth all the way down to the skin. To prevent a flea's jumping from the comb, zap it with a cotton swab dipped in petroleum jelly before drowning it. Since they may be carrying harmful organisms, don't crush the fleas with your fingernails.

Light-and-Water Trap

African villagers put a lighted candle in a dish of water which they set in the middle of their huts to attract fleas. You can modernize the idea with a flexible-necked lamp. Set the lamp and a shallow dish filled with water and detergent near an infestation. The detergent will eliminate the surface tension that could turn the water into a flea trampoline. Leave the light on all night for at least a month, says Jo Frohbieter-Mueller, writing in *Mother Earth News*. The fleas will jump toward the light's warmth and fall into the water.

Bathing

Regular, frequent bathing in summer can control fleas on dogs and that rare jewel, a cat who doesn't mind getting wet. Since fleas drown quickly, an insecticidal soap is unnecessary.

If your pet objects strenuously to getting wet, try a sponge bath with denatured alcohol, which may be mixed with vinegar; or sponge your animal with a strong brew of wormwood tea. Wormwood leaves are available from commercial herbalists.

The insecticides that promise so much can prove less than adequate in controlling fleas, even if you don't mind the exposure to the toxic products. People have had their houses and yards treated, their animals dipped in organophosphates, but the pests soon come back.

Diligence

When you find the combination of repellents, baths, combing, and environmental controls that works best for you and your pets, these should do at least as well as the chemicals. However, they are not one-shot solutions; you must apply or use them regularly and frequently in warm weather.

For the long haul, look into improving your pet's nutrition. Some veterinarians who promote high nutrition for dogs and cats claim it can increase an animal's resistance to fleas.

If you really can't tolerate any fleas at all, you should find another home for your pet. Or you might consider adopting a horse. Horses don't have fleas.

Ticks

For city dwellers and suburbanites who have dogs, ticks are not much of a problem. Cats, possibly because they lick themselves so frequently and thoroughly, also rarely get ticks. And you can have a dog for many years and never encounter the pest. But, as with all other forms of wild-life, moving closer to wilderness areas, or acquiring vacation homes in mountains and forests, we're most likely to meet up with this relative of spiders, mites, and scorpions.

Fortunately this creature is not nearly so invasive as fleas. Except for one species, the brown dog tick, it is normally found only on grass or bushes, and does not become established in homes or kennels.

Hazards of Ticks

Some species are dangerous, causing paralysis or infecting their host with Rocky Mountain spotted fever, tularemia, or relapsing fever. The bites of all ticks can be an uncomfortable nuisance and, if neglected, may lead to serious complications.

A dog infested with ticks can be greatly weakened by loss of blood. The adult parasite is most often found on the dog's ears and neck and between its toes. Larvae and nymphs (the young adults) settle on the long hairs along the animal's back.

The American dog tick transmits Rocky Mountain spotted fever; the brown dog tick does not. If you're not sure which species is feeding on your dog, after looking at the illustrations, take a few specimens in a jar to

Figure 8.2. Brown dog tick (left); and American dog tick (right). ("*Common Ticks Affecting Dogs.*" Division of Agricultural Sciences, University of California. Leaflet 2525, June 1978.)

your local farm bureau or health agency for positive identification. It's safest to handle the specimens with a tweezers to avoid contact with any disease organisms they may be carrying.

Dealing with Ticks

As with many pests, prevention is your best weapon. On returning from a walk with your dog in a rural area, check the animal for ticks. When the parasites first attach themselves they're fairly easy to remove and have had little chance to pass disease organisms to the dog.

The Centers for Disease Control recommend rodent-proofing your home, especially if it's a mountain cabin, to keep out tick-bearing rats, mice, squirrels, and the like. (See Chapter 5.)

Keep your shrubs and lawns closely trimmed so that the parasite and their rodent hosts find no hiding places near your home.

Removing Ticks Safely

To remove ticks safely, treat them as you would any large splinter, using a pair of tweezers and an antiseptic. Try not to crush the tick while pulling at it. It is most important that the animal's tiny head not be broken off and left in the wound. If this happens, ulcers, infection, even blood poisoning can result.

Substances That Make a Tick Let Go

chloroform	Vaseline
isopropyl alcohol	nail polish
ether	lighted cigarette
acetone	adhesive tape

Use a slow, steady pull, say Harry Pratt and Kent Littig of the Centers for Disease Control, to minimize the chance of complications from detached mouthparts. The entomologists continue:

> "There is no certain way to make a tick detach its mouthparts. A drop of chloroform, isopropyl alcohol, ether, acetone, or Vaseline or fingernail polish rubbed over it may help remove the tick. Several minutes to a half hour later, when the tick has withdrawn its mouthparts, it can be removed with less damage to the skin. Heat from a lighted cigarette* sometimes causes it to release attachment. An antiseptic should always be applied to tick bites just as to other wounds. If the hands have touched the tick during removal, wash them thoroughly with soap and water since the tick secretions may be infective."

Entomologist Karl von Frisch recommends applying fresh adhesive tape if the tick is not yet very swollen. Perhaps the chemicals in the adhesive irritate the pest or perhaps the tape simply suffocates it. At any rate, when you remove the tape the next day, the tick will come away with it.

* Don't burn the tick with the cigarette; just hold it near enough so the animal feels the heat.

---- CHAPTER 9 ----

The Unmentionables: Lice and Bedbugs

A few years ago, soon after fall classes started at an Ivy League college, a student asked a classmate with whom she'd just spent the summer working at a children's camp to come to her room. With the door closed, she asked, "Do you know lice when you see them? I think I've got them."

Lifting locks of clean short brown hair, the friend spotted one or two tiny grayish-white creatures scurrying away from the light.

"I think you do too," she said, "and I think these are nits," she added, pulling off little yellowish ovals from a few individual hairs.

This unlucky young woman spent the next 2 weeks in the college infirmary, her head frequently wrapped in a towel.

Though usually linked to personal filth and dirty surroundings, lice are really democratic, and will readily move from one individual to another no matter how fastidious the new host, nor how well placed on the social ladder.

Bedbugs are equally undiscriminating. Not long ago a family whose home boasted a grand piano and oriental rugs bought an almost-new sofa bed from acquaintances. The first night the family's son slept on it he was badly bitten and the next morning had itching welts all over his body. The bed was quickly thrown out.

It's easy for the unwary to pick up either of these nuisances and very difficult to get rid of them. First, let's deal with the more dangerous of the two—lice.

Lice

Lice, like mosquitoes and fleas, are carriers of diseases that have killed millions. They are the inevitable accompaniment of war and other catastrophes when people can't change their clothes or bathe frequently.

Louse-borne typhus (more deadly than that carried by fleas) has at times been more important than military skill and strength in deciding a war's outcome. In part because typhus decimated the French army in 1528, the Spanish defeated them, and Spain dominated Europe for the rest of the century.

During World War I typhus and trench fever, also louse-borne, killed German and Allied soldiers evenhandedly. In World War II typhus raged in North Africa, the Balkans, Russia, Italy, and the Nazi concentration camps, killing soldiers and civilians alike.

But that war brought a breakthrough in control of typhus and the parasite that carries it. DDT, newly invented in Switzerland, was dusted on people during an incipient typhus epidemic in Naples and the disease was halted. For the first time in history, a typhus outbreak was stopped in winter. Today the disease can be cured by antibiotics. But the lice live on.

Body, hair, and pubic lice (crabs) are presently on the rise. There are several reasons, according to Benjamin Keh, foremost U.S. expert on these parasites.

Ignorance. This generation has had little experience with lice. Since they don't recognize the early signs of lousiness, or pediculosis, they can't take the necessary steps to control it.

DDT Curbs. Restrictions on the use of DDT because of its long-term harm to the environment is another cause.

Insect Resistance. With that remarkable ability of the many-legged to fight back, lice have become increasingly resistant to other insecticides.

Lax Lifestyle and Lower Personal Standards. These standards took hold in the sixties and have lingered into the eighties characterized by long, unwashed hair, a tendency not to bathe or change clothing often, and greater sexual permissiveness. In addition, the thousands of homeless wandering our streets during the hard times of the early 1980s have probably contributed to the upsurge.

Nature of Lice

Characteristics of Human Lice

Appearance

Lice are wingless, round, flat insects from 1 to 4 millimeters long. They're grayish-white (crab lice may have a pinkish tinge) and have three sets of short, stout legs. One or more of the legs will have a claw, well suited to grasping a human hair. Pubic lice are shorter and broader and have two pairs of heavy claws.

Habitat

Head and pubic lice spend their entire lives on their host. Body lice, by contrast, hide in the person's clothing when not feeding.

Reproduction

Because humans are warm, lice breed all year long. Crabs and head lice cement their eggs to body or head hair; body lice to the fibers of underclothing, especially along the seams and at the neck, shoulder, armpit, waist, and trouser crotch. With the necessary warmth, eggs hatch in about a week. Temperatures above 100° and below 75°F either greatly reduce hatching or prevent it completely. The insect lives about 3 weeks.

Need for Warmth

The parasite cannot survive if the host's body temperature drops due to death or rises due to fever. It, however, can readily change hosts if there is close personal contact. Off the person, all life stages of lice die within 30 days regardless of temperature.

Lice Don't Jump or Fly

Lice move very rapidly especially when exposed to light. Contrary to what you may hear or observe, they neither jump nor fly. However, static electricity in a comb can expel the creature suddenly, giving the impression of a leap or flight.

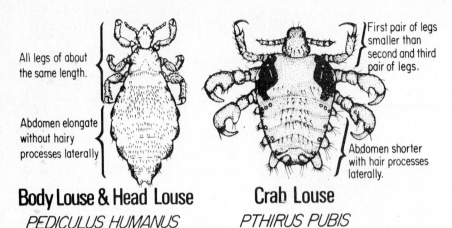

All legs of about the same length.

First pair of legs smaller than second and third pair of legs.

Abdomen elongate without hairy processes laterally

Abdomen shorter with hair processes laterally.

Body Louse & Head Louse
PEDICULUS HUMANUS

Crab Louse
PTHIRUS PUBIS

Figure 9.1. Body louse and head louse (left); crab louse (right). (*Lice of Public Health Importance and Their Control*. Atlanta: Centers for Disease Control. Homestudy Course 3013-G, Manual 7-B, 1982.)

How People Pick Up Lice

The parasite can move directly from one person to another.

Body lice are picked up by direct contact with an infested person's body hairs or clothing. They also infest bedding used frequently by an infested person and can move to the bed's next occupant.

Head lice are transmitted by shared combs and brushes, wigs, hats, athletic headgear, and towels.

Crab lice are most often passed during sexual contact, but can also be picked up from public toilets.

You cannot get lice from an animal or insect. These parasites are passed only from one person to another through direct contact or shared clothing, bed linens, towels, and upholstery used by an infested individual.

You don't *have* to be personally filthy or live in squalid surroundings to catch lice, but it helps. Once you've got lice, filth will guarantee their increase and crowded conditions will spread them.

Body lice peak in winter in cold climates, when people wear several layers of clothing. Typhus, which is carried primarily by body lice, is endemic in such regions in the cold months. On the other hand, you can be finicky about your person, your friends, and your environment and still contract head lice or crabs, although such bad luck is more often the result of poor personal hygiene or loving contact with someone who has them. School-age children, for whom bathing and shampooing are apt to be low priorities, are the most frequent carriers of head lice.

Since lice do not easily drown, soap and water at temperatures we can stand don't get rid of them. That's why in many parts of the world, people who continually wash their bodies may still be lousy.

Signs of Lice

The classic symptom of long-term lousiness is scarred, hardened, or pigmented skin; but long before getting to that stage, one can easily spot telltale signs of specific species.

Body lice

Dirty brown patches of their droppings around the victim's armpits. Scabs, excessive scratching, and inflamed, irritated skin. Lice seen on the outer clothing signify many more on the inner garments.

Head lice

Brown fecal material on shoulders and back, easily seen on light-colored clothing. Louse eggs, or nits, along the hairline behind the ears. More head scratching then usual, difficulty sleeping, and irritability. Probably the most common species in the United States, head lice usually infest school-age children.

Crabs

Dark spots on the victim's underwear or around the armpits. Itching, at times severe. Occasional painless blue spots on the skin, not seen with head or body lice. On infants, who sometimes acquire crabs from their care givers, the parasites can live on the eyelashes, eyebrows, and along the hair line.

Shut Lice Out, Wipe Lice Out

As with most parasites, both human and animal, it's simpler to prevent an infestation than to clean one up.

Body Lice

- Bathe and change your outer clothes frequently, and your underwear daily.
- Don't go without underwear. Since it's changed and laundered often, just wearing underwear regularly is a good preventive.
- Avoid borrowed clothing; you could be borrowing lice.
- When buying used garments, either from an acquaintance or a thrift shop, launder them or have them dry cleaned before wearing them. Or store the items for 4 weeks in a plastic bag to kill any lice and their young.
- Avoid sleeping in beds used by people you think may be infested. If you must use such a bed, make sure linens and blankets are fresh.
- Inspect your body and clothing frequently if you think there's a chance of infestation.
- Avoid contact with infested people, especially at night, when the parasite is more active.

Getting Rid of Body Lice without Chemicals

Noninsecticidal control of body lice is simple.

1. **A Special soak.**   Lice on the person can be killed by applying an emulsion of two parts kerosene, one part soap chips, and four parts water. Work up a lather and let it remain on the body for 15 minutes before rinsing. This method is recommended by the University of California Cooperative Extension Service. Need I add, "Don't smoke"?

2. **Properly cleaned clothing.**   A person infested with body lice who changes frequently to properly laundered clothing will eventually get rid of the parasite without any other treatment. Proper laundering means agitating washable garments in *hot* water, at least 140°F, for 20 minutes. If circumstances are such that you can't wash them, store the items in a sealed plastic bag for 30 days. This will kill any lice in the garments.

   The University of California Cooperative Extension Service, recommends washing clothing and bedding with a 5 percent creosol solution, or soaking them for 30 minutes in a 2 percent solution of the substance.

3. **Dry cleaning.**   Woolen clothing belonging to an infested individual should be dry-cleaned. Cleaning solvent is lethal to lice, as is the heat of pressing. Put the suspect garments in a tightly closed bag before taking them to the shop. Not every cleaner will accept infested clothing, especially those who don't operate their own plants, so check around first. The Centers for Disease Control say that just pressing woolen garments along the seams gives good control since this is where body lice lay their eggs.

Head Lice

According to Dr. Paul Ehrlich, assistant professor of pediatrics at the New York University Medical Center, more than 6 million schoolchildren catch head lice every year. The heaviest time is in the first 2 months of school. Says Dr. Ehrlich:

> "Usually, lice are spread by children returning from summer vacations—unaware that they are infested. By the time the insect's eggs are hatching, children may have shared combs, brushes, and clothing, spreading the problem. The problem spreads through school coat racks, lockers, and fabric-upholstered furniture. Under these conditions the lice move from one source to another so quickly the spread is difficult to contain and schoolwide epidemics are the result."

In addition, adults often assume that school-age children are able to care for their own hair and don't supervise shampooing. The youngsters themselves don't recognize the real nature of their problem. They just scratch.

Lice are no respecters of wealth or cleanliness. Actually, according to dermatologist Lawrence C. Parish of Philadelphia's Jefferson Medical College, they prefer clean scalps. Even youngsters in exclusive schools aren't safe.

Preventing Head Lice

Whether you're rich or not so rich, preventing head lice is a matter of a few simple procedures consistently carried out.

Braids or Very Short Hair.   As early as 1913 it was recommended that *girls* braid their hair before going to school and unbraid and comb it when they come home. Braiding keeps the hair from contact with that of other children, preventing it from flying about and possibly picking up a few lice from a playmate. Another common practice was cutting *boys'* hair very close. It's still a good precaution.

Assigned Clothing Hooks in School.   Head lice spread quickly when children store their clothing haphazardly at school. Youngsters with assigned hooks and/or individual lockers have fewer cases of pediculosis than those who share clothing storage. If you have school-age children, insist that school authorities provide individual storage for students' clothing.

Coconut Oil Shampoo.   Another sound preventive is regular shampooing with a coconut-oil based shampoo. Coconut oil is the source of dodecyl alcohol, lethal to adult lice and some eggs. Green soap, once the standard shampoo at summer camps and other institutions, is based on coconut oil.

Identifying Head Lice

An individual with tiny yellowish globules on the hair, especially behind the ears at the hair line, but with no visible lice, may or may not have an infestation. The globules could be hair spray or dandruff. To be certain, try to slide them along the hair shaft. Dandruff and hair spray will slide right off but nits, which are cemented firmly in place, will not. Unhatched eggs will be $\frac{1}{4}$-inch or less from the scalp. Any further down the shaft have already hatched.

A Safe Alternative: Shampoo, Soak, and Comb

The quickest, cheapest safe control for head lice is a drastic hair cut, $\frac{1}{4}$-inch or shorter, or shaving the head, both of which could bring unbearable ridicule on a child. There is another, less traumatic alternative, one proven effective for generations. It's a combination of shampooing, soaking, and fine combing. Pediculosis expert Benjamin Keh has adapted instructions of the manufacturer of the Derbac nit-removing comb for eliminating lice. He recommends a shampoo followed by a soak in the lather; the Centers for Disease Control suggest that the soak come first. Benjamin Keh's method follows.

1.   Wet hair thoroughly with water that is quite warm.
2.   Apply a soap shampoo, [preferably coconut-oil based] and work to a thick lather. Rubbing scalp and hair thoroughly, start from back of neck, ears, and forehead and work toward center of

head. Be sure entire head and all the hair are covered with shampoo.

3. Brush this lather off and rinse with water, again quite warm.

4. Repeat lathering and rubbing, this time leaving thick suds on the hair. Tie a towel around the lathered head and leave on for a half hour.

5. Remove towel. Hair will be soapy. *Leave it this way.*

6. Comb hair with regular comb to remove tangles.

7. Separate 1-inch strand of hair. Using the Derbac comb, hold it with beveled side toward the head. Lifting the strand of hair away from the head, comb the strand slowly and repeatedly from scalp to end of hair until every nit is removed. (If comb misses any nits, remove them with your fingernails and drop them into boiling water.)

8. Pin clean strand out of the way. Start the next strand and comb in the same way. Continue combing strand by strand until entire head is free of nits.

9. *Important:* If hair dries during combing, wet with water. Wipe nits off comb frequently with cleansing tissues. It usually takes 2 hours of constant combing to remove all nits, longer if hair is thick and long.

10. After removing all nits, lather hair again, rinse, and dry.

11. When hair is thoroughly dry, inspect entire head for any missed nits and remove them.

12. Following the shampoo and combing, immediately soak Derbac comb in hot, soapy ammonia water for 15 minutes. Then scrub teeth with stiff nail brush. Remove dirt lodged between teeth by pulling number 8 thread through the openings. Do not leave any dirt on comb teeth. Boil the towels used in shampooing and the brush.

If your pharmacy cannot supply you with a Derbac comb, contact the manufacturer:

Cereal Soaps Company
Division Johanson Manufacturing Corp.
Box 329
Boonton, NJ 07005

Because there is no known substance that will dissolve the cement gluing louse eggs to the hair without damaging either hair or scalp (vinegar, contrary to popular conception, will not do the job), the method described above is the safest, most effective way to deal with an infestation of head lice.

If your child's school is suffering an outbreak of head lice, you must inspect and comb the youngster's head each day right after school. A comb warm enough to be pleasantly hot to the hand will rout the lice from their hideouts among the hairs, making them easier to remove. Have a hand-held hair dryer nearby to warm the comb from time to time as you work. Be sure to drop *all stages* of the organism that you catch into boiling water to kill them and their disease germs. Don't crush them with your fingernails as this will release their harmful germs onto you.

During a school epidemic of pediculosis you should also check every other member of the family since the parasite moves very quickly from person to person in close quarters. If you're inclined to shrug the whole problem off, remember that a neglected case of head lice can produce a foul mass on the scalp, one in which a fungus readily develops.

Since lice can be transmitted through upholstery, vacuum sofas and chairs every day until the infestation is over. Take special care to launder all clothing frequently and dry-clean nonwashables. As with body lice, you can also store garments in plastic bags for 4 weeks.

Lindane: A Drastic Measure

Under various commercial names, lindane is the insecticide most widely used to clean up head lice. Until 1983 the federal government severely restricted its use against organisms other than head lice, and with good reason. Lindane can be absorbed through the skin, and has caused convulsions, seizures, and cancers in laboratory animals. One child is reported to have died from lindane poisoning after treatment for head lice, and another had a seizure. The insecticide has also been implicated in several cases of a grave form of anemia.

Admittedly lindane has been used many times without any reports of ill effects, but that will make little difference to the parent whose child is the exception. In addition, lice are becoming resistant to the chemical so exposing your child to this powerful substance may not clear up the infestation. Like many other chemicals we've been using so freely over the past few decades, lindane's long-term effects are not yet completely understood.

Only as a last resort should you use a pesticidal shampoo. The safest are those based on pyrethrins, at a strength not less than 0.3 percent, according to William and Helga Olkowski of the Center for the Integration of Applied Science.

Remember this old saying, "It's no disgrace to have lice, but it is to keep them."

Note: If your treatments at home do not control the infestation, take your child to a doctor. You should also seek medical advice for anyone who gets a severe reaction to louse bites.

Crabs

Like venereal disease, the occurrence of crabs, or pubic lice, has increased greatly with the sexual revolution. Even people with a circumspect sex life, or no sex life at all, can attract these unpleasant, embarrassing little pests since they can infest public toilet seats, beds, and towels and clothes in gymnasium locker rooms or wherever many different people's clothes are crowded together. They can also be acquired from people's stray body hairs.

Although usually living in the armpit, pubic, and perianal areas, crabs can also infest beards, moustaches, eyebrows, eyelashes, and chest hair—wherever hairs are spaced fairly wide apart. On infants, who can pick them up from their care givers, they are found on the eyelashes and eyebrows and along the hairline.

Preventing Crabs

1. If no paper cover is provided in a restroom, turn the toilet seat up for a few seconds before using it so any lurking parasite slides off.

2. Don't share gym towels or lockers.

3. Make sure any strange bed you sleep in has fresh linens.

Treating Crabs

The simplest treatment for crabs, say the Centers for Disease Control, is the same as for head lice: Shave the affected areas very close, thus eliminating all stages of the pest. *Mercuric ointment, used for generations, is neither effective nor safe.* Its value as an insecticide is low and it carries serious risk of poisoning.

In young children and infants, apply 0.25 percent physostigmine eye

ointment with a cotton swab to eyelashes, eyebrows, and along the hairline. The ointment can be bought only with a doctor's prescription.

To guard against reinfestation, boil all underwear, bedding, and towels
for the entire household.

Summary of Nonchemical Steps in Controlling Lice

Body Lice

- Apply for 15 minutes emulsion lather of two parts kerosene, one
  part soap chips, four parts water.
- Launder all linens and washable clothing in 140-degree water for 20
  minutes. Boiling and use of creosol optional.

Head Lice

- Cut hair *very* short, $\frac{1}{4}$-inch or less, or
- Shampoo thoroughly followed by
- Combing wet hair, 1-inch strand at a time, with very fine-toothed
  comb, or
- Handpick lice and eggs after shampoo.
- Repeat handpicking and/or combing for 12 days.
- Inspect and fine-comb daily thereafter until school epidemic is
  over.

Crabs

- Shave all body hair
- Boil all underwear and linens used by infested person.
- Boil household linens after use.

Bedbugs

"We get calls from people in Beverly Hills," says Corinne Ray, director of the Los Angeles Poison Information Center, "They ask, 'What
should we do about all the spiders in our bedroom? We're being bitten all
over.' "

Spiders only bite when they feel threatened, she tells her callers, and
then only once. "You probably have bedbugs," she adds. The callers
abruptly hang up.

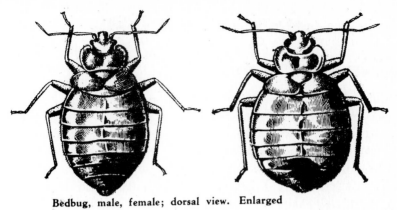

Bedbug, male, female; dorsal view. **Enlarged**

Figure 9.2. Bedbugs. (Hugo Hartnack, *202 Common Household Pests of North America*. Chicago: Hartnack, 1939.)

Few people like to admit they have bedbugs because the pest is so often associated with dirty homes and poor hygiene. But a generation or so ago they were quite common, especially tormenting travelers even in "respectable" hotels.

There was a great surge of bedbugs in northern Europe in the 1930s, brought on, it is believed, by the rapid adoption of central heating. The problem was so bad in Sweden that a number of cities seriously considered building special hotels to house people whose homes were being fumigated. Tents, used in summer, were out of the question in Sweden's cold winters.

Around the same time about 4 million Londoners were living in bedbug-infested homes. Fumigation of furniture was almost compulsory and the homes themselves were thoroughly treated between occupants.

Bedbugs are still a problem under primitive or unsanitary conditions. And when people move from infested housing to clean accommodations, they easily carry the pests with them. Even well-kept university housing, according to Professor Walter Ebeling, is not exempt.

The insects invade otherwise clean homes from various sources. Secondhand furniture is the most frequent origin, but the pest can come from bat roosts, swallows' nests, chicken and pigeon coops. Other sources include ill-kept theaters, furniture retrieved from storage, poorly maintained upholstery on public transportation, infested clothing laid on a bed, and moving vans that have carried infested furniture. If an infested apartment is vacated and remains empty, the pests will spread, in time, to nearby occupied quarters.

Nature of Bedbugs

Dangers from Bedbugs

Although they have not proven this beyond doubt, scientists say that bedbugs may transmit plague, relapsing fever, tularemia, and Q fever. This last, related to Rocky Mountain spotted fever, was first identified in Queensland, Australia, hence the odd name. Even without the possibility of disease, the bedbug is a most unpleasant visitor. Its bite leaves hivelike welts that itch intensely. Like fleas, bedbugs pierce the skin to suck their host's blood. Also like fleas, their saliva causes more discomfort than the bite. With extremely sensitive people, the swelling caused by the saliva can be severe.

Bedbug Habitats

Once in a home, bedbugs concentrate on rough, dry, partially, or completely dark surfaces. They prefer wood or paper, where their excrement and eggs soon become visible. They like to lay their eggs, firmly cemented in place by a sticky secretion, behind loose wallpaper, in wall and floor cracks, nail holes, light switch boxes, and on door and window frames. They will hide in a wood bed frame but you may not be safe even if yours is metal because the insect readily travels from its hiding place to find a blood meal, even if it's in another room.

The pest gives off a characteristic smell, rather like very sweet raspberries only, say some, if they've been fed on by stinkbugs.

A Hardy Breed

Bedbugs, also called mahogany flats, redcoats, chinch bugs, and wall lice, by whatever name, are a hardy breed. Four to five millimeters long, oval, and flat when hungry, the creature becomes plump and elongated after a feeding. Under normal conditions it can live as long as a year, but starvation and cold weather prolong its life. Without food and at 60°F or below the adult bedbug enters semihibernation and can overwinter in an unheated building. Its eggs and newly hatched young, however, die within a month at subfreezing temperatures and so do not survive winter in an unheated structure.

Shut Bedbugs Out

Prevention the Best Control

Bedbugs are no different from other household pests in that it's simpler to prevent them than to clear them out. Here are some preventive steps:

- Inspect any used furniture before buying it, including beds, upholstered pieces, rough-textured draperies.
- Patch all wall cracks, nail holes, and crevices with caulking compound or latex paint. Make sure all window and door frames are completely sealed and free of cracks.
- Paste down any loosened wallpaper securely.
- Fill all floor cracks. A fresh coat of varnish on a wooden floor helps.
- Caulk all cracks and spaces behind baseboards.
- See that the paper backing on all framed pictures is securely pasted down.
- Make sure your house's foundation is crack-free.
- Attic vents should be screened so bats and birds can't get in to roost.
- Pipes entering the house should be fitted with flanges.
- Destroy nearby bird nests and discourage further nest building.

An Early Method of Control

In Austria in the 1930s bedbug-proof construction was considered the surest control. This included coating walls from the floor to three-quarters of the height of the room with slick oil-based paint. Instead of a conventional baseboard, hollow molding, concave to the floor, finished the room. There's no reason why this wouldn't work today, although shiny walls may not be your idea of good decor.

Wipe Bedbugs Out

Getting rid of bedbugs quickly without chemicals is difficult. But before resorting to hazardous substances, try some other means first.

- For temporary control, the Center for the Integration of Applied Science recommends smearing petroleum jelly around the legs of your bed (assuming the insect isn't hiding in the bedframe but just migrates there from the wall).

- Send your mattress and pillows out to be fumigated.

- Vacuum your room and its accessories (such as picture frames) thoroughly.

- Various forms of heat can give good control: Focus a high-intensity lamp on mattress seams and buttons and on wall cracks for a quick kill.

- Expose mattresses and bed frames to pressure-generated steam.

- Wash bed linens and blankets in *hot* water.

- Heat your house to 115°F. All bedbug stages die above 113°F.

The Safest Chemical Controls

The safest chemical to use against bedbugs is pyrethrum dust, laid down in all cracks and crevices likely to be harborages.

If the insects are coming from the attic or wall voids, Walter Ebeling suggests applying silica gel Dri-die 67 to these areas. Because of its irritating qualities, be sure to wear goggles and a respirator when applying it.

If You Must Use Chemicals . . .

People get even more uptight about bedbugs than they do about cockroachs. If you feel you absolutely *must* use a strong chemical insecticide, this warning from the Centers for Disease Control should save you grief:

"Special care should be exercised in the treatment of mattresses and upholstery so that only a light application is made. *Under no circumstances should mattresses be soaked with spray. Infant bedding, including the crib, should not be treated. With dichlorvos or trichlorfon, spray only tufts and seams of mattresses, and air until dry—at least 4 or 8 hours, respectively—before reuse. If infesta*tion persists, re-treat at not less than 2-week intervals."

Natural Predators

There has been much interest recently in biological control of insect pests, with some notable successes. Bedbugs are not one of them. Their predators are just as undesirable as they are. Who'd want to introduce centipedes, straw itch mites, pharoah ants, and cockroaches into their bedrooms just to get rid of bedbugs?

PART FOUR

PESTS OF PROPERTY

CHAPTER 10

Clothes Moths, Carpet Beetles, Silverfish

Clothes moths, carpet beetles, and silverfish can rob you of as much as a clever burglar. Unlike burglary, however, it's next to impossible to get insurance against insect damage. Your best protection is being aware of what these pests need and eliminating those conditions.

Moths and carpet beetles destroy fabrics of animal origin. Silverfish, which eat carbohydrates, can ruin your library or family papers. They also attack textiles of vegetable origin like cotton and linen and seem especially fond of rayon. Since the first two insects wreak much the same kind of havoc, we'll consider them together.

Clothes Moths and Carpet Beetles

In nature, clothes moths and carpet beetles are useful, eating remains of dead animals that other scavengers leave behind—hair and feathers, claws and nails, horns and hoofs (substances made mostly of keratin, a protein these insects can digest). By using hair and other animal products in our clothing and home furnishings, we've created an ideal environment for fabric pests.

Clothes moths and carpet beetles can attack a new sports jacket, a treasured Persian rug, a favorite old bathrobe, the horsehair on a violin bow as well as sable-bristled art brushes—they do not discriminate be-

159

tween old and new. Fabrics and carpets imported from a country where products aren't mothproofed during manufacture are also susceptible. And even items so treated lose this protection with cleaning and over time. Clothes moths thrive in warmer regions; carpet beetles in cooler regions.

Fortunately, both insects are less of a nuisance than they were years ago since today so many of our home furnishings and clothes are made of synthetics that the pests don't favor. Listed below are the many materials that are liable to damage from clothes moths and carpet beetles.

Items of Wool

clothing	rugs
blankets	felt insulation and weatherstripping
upholstery	piano felts

Items of Fur, Hair, or Feathers

fur coats	upholstery
artists' brushes	pillows
bows for stringed instruments	

Other Substances

paper	tobacco
straw	spices
cotton	hemp
rayon	animal skins

Substances Attracting Moths and Carpet Beetles to *Any* Fabric

fruit juice	grease
beer	urine
milk	perspiration
gravy	fungus spores

Carpet Beetles Also Attack

milk powder	cayenne pepper
casein	beans, peas, corn, wheat, rice
books	seed products

In both species, it is the larvae who are the culprits—the adults do no harm.

Why Not Chemical Insecticides?

The chemicals widely sold to protect furniture and clothing against insect damage can expose you and your family to unnecessary health risks. Curious children often pop antimoth nuggets into their mouths, thinking they're marshmallows or rock sugar. If your winter clothes, generously sprinkled with chemical antimoth crystals, are stored in a bedroom, sleepers are breathing in the fumes. And if you don't have your sprayed garments dry-cleaned before wearing them again, you could be absorbing chemicals through your skin. Other protective measures work just as well as the chemicals and with much less risk.

Signs of Fabric-Eating Insects

Finding holes in your garments or furniture may be your first tip-off to the presence of fiber chewers. Knowing which pest it is can help in dealing with the problem. Carpet beetle holes are spaced rather far apart. The larvae leave their excrement nearby, and it's usually the color of the fabric being eaten. You may also find their cast-off skins, which are pale and rather bristly.

Moth damage is apt to be more localized since these larvae are more sluggish than those of the carpet beetle.

The first reliable sign of clothes moths is usually a damaged fabric or the larvae's webbing or cases. The occasional moth wandering around inside your home is most likely another species, possibly a stray from outside or a grain moth. Clothes moths are not drawn to light and hide when disturbed; females heavy with eggs are weak flyers.

By contrast, carpet beetles clustering around your windows are a signal for firm action. They are reproductive adults, with eggs ready to hatch. You may already have wondered about their odd-looking larvae wandering from room to room.

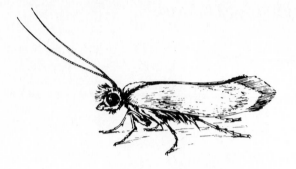

THE CLOTHES MOTH

Figure 10.1. The clothes moth. (Reprinted with permission from Karl von Frisch. *Ten Little Housemates*. Oxford, England: Pergamon Press, 1960.)

Nature of the Clothes Moth

Clothes Moths: Appearance and Habits

In the United States two species of clothes moths do most of the damage. The *webbing clothes moth* is the most common, the *casemaking clothes moth* somewhat less so. Equally destructive species are found in other parts of the world. The chart on p. 164 shows how casemaking and webbing clothes moths differ.

In addition to its well-known taste for wool, the larva will also attack paper, straw, cotton, or rayon, using their fibers to spin its case or mat. The eggs can hatch under water and the larvae survive there for 26 hours. One species, when fully grown, can gorge on mothproofed fabrics with no ill effect. The average larva will eat eleven or twelve times its weight in wool. It can starve for more than 8 months and still become a fertile adult. This tiny creature is tough!

Sources of Clothes Moth Infestations

Any item that uses, gathers, or consists of hair or bird feathers can attract moths. This includes:

Hair and feathers of dead rodents and birds

Insect remains

Felt insulation and weatherstripping (found in older homes)

Infrequently changed vacuum cleaner dust bags

Piano felts

A felt pad under a typewriter

Abandoned stuffed toys

A rarely used down quilt

A pillow with a torn cover

All can support colonies of moths, whose growing populations will migrate in search of more food. Regular checks and frequent cleaning, especially in attics or basements (often the site of dead rodents or insects and unused wool items) will give you good moth protection.

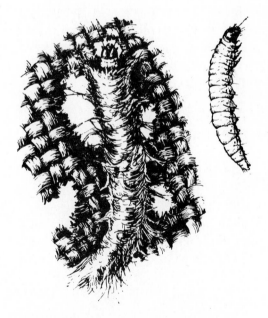

Figure 10.2. Caterpillar of a moth in its case (left); side view of a caterpillar removed from its case (right). Magnified about four times. (Reprinted with permission from Karl von Frisch. *Ten Little Housemates*. Oxford, England: Pergamon Press, 1960.)

	Webbing Moth	*Casemaking Moth*
Appearance	golden buff color	drab buff color
	satiny wings	vague dark spots on wings
	brownish head hairs	light-colored head hairs
	about ¼-inch long	same
	½-inch wingspread	same
Abilities	without eggs, can fly considerable distances	same
	penetrates very narrow cracks	same
Breeding habits	eggs laid on rough textiles	same
	summer hatching time: 4 to 10 days	same
	winter hatching time: up to 3 weeks	same
Larvae	⅛-inch long	same
	pearly white	same
	stage lasts 50 to 90 days	same
	spin netting either as flat mat or feeding tube	spin case and drag it about
	prefer inside folds of clothing: pleats, pockets, collars, cuffs	prefer hair, feathers, tobacco, spices, hemp, animal skins

Nature of the Carpet Beetle

Carpet beetles, except for the black common carpet beetle, are beautiful and colorful, shaped like the helpful lady beetle. If it didn't move, the

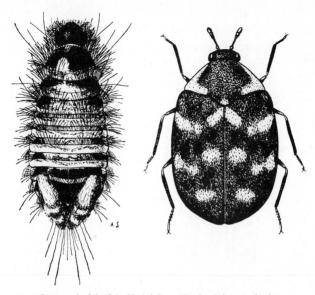

Larva and adult of the Varied Carpet Beetle, *Anthrenus verbasci.*

Figure 10.3. Larva and adult of the varied carpet beetle. Magnified about seventeen times. (*Carpet Beetles.* London: British Museum (Natural History), Economic Leaflet No. 8, 1967.)

furniture carpet beetle, mottled black, white, and golden brown, could be mistaken for a semiprecious stone. The varied carpet beetle is a little less showy, running more to subdued grays and blacks. (See Figure 10.3.)

But don't be deceived by good looks. These insects are the most destructive fabric pests in the world—and the most difficult to control. Their larvae are relatively resistant to pesticides, reason enough to use nonchemical controls.

Adult carpet beetles, also called buffalo beetles, are 2 to 5 millimeters long. Outdoors they feed on the pollen of flowers, especially white or cream-colored ones. Indoors they fly to window light and, unlike the shy clothes moth, roam freely about a house. Their sturdy larvae, up to 8 millimeters long and resembling stubby, bristly carrots, move very fast.

Carpet Beetle Damage

These larvae prefer napped fabrics, munching the short fibers from the free end to the base fabric in sharply limited spots. They also eat smooth textiles. In furs, the grubs lie parallel to the hairs, almost hidden, their heads facing the skin and their bristly rears toward the outer surface. They burrow through packaging material to get at food, clearing the way for other pests. One odd species traces with its chewings the cracks in the floor beneath a carpet.

Carpet Beetles' Habits

The female carpet beetle lays her eggs in dark protected crannies, behind baseboards, in floor cracks, and inside hot-air ducts. The eggs hatch in 1 to 2 weeks. From egg to adult may take 3 years, but in a centrally heated building 7 months is the usual span. Because these larvae are so active, you may not find any on the materials they've damaged.

Sources of Carpet Beetle Infestations

The staging area for an invasion of these insects may be hard to find. In addition to baseboards, radiators, and other places where lint and food crumbs collect, the insects could be coming from heating and air-conditioning ductwork, wall voids, attics, and subflooring. Nests of birds, wasps, or bees near your house, as well as light-colored flowers could also be the source.

According to entomologist Walter Ebeling, they've even been found feeding on telephone cable insulation. If they're in an attic, their cast-off skins can drop through openings around beams and light fixtures, or through the perforations in acoustical tile. An easygoing homemaker could find that cereal cartons have attracted a carpet beetle gang, who soon invade the rest of the house.

Shut Carpet Beetles and Clothes Moths Out

Preventing insect fabric damage is much cheaper than replacing clothing or furniture. To work, however, prevention must be thorough.

Outdoors Control

1. Remove bird nests (old or new), abandoned spider webs, and wasp nests hanging under the eaves.
2. Block up all attic openings to keep out rodents.

3. Replace any felt insulation around pipes or under the roof with nonorganic material.

4. Seal all openings around pipes entering the house.

5. Avoid planting white flowers in your garden.

6. When bringing cut flowers inside, check them for feeding carpet beetles.

Indoor Control

1. Caulk any spaces between baseboards and walls and floors.

2. Check your basement and attic and clean out any dead birds, insects, or rodents.

3. Clean closets and dresser drawers regularly.

4. Keep little-used parts of your house free of dust and lint.

5. Rearrange furniture occasionally to expose carpets to light since keratin eaters prefer dark, secluded corners.

6. Move heavy furniture once a month so you can vacuum under it.

7. Vacuum along baseboards; behind radiators; draperies, valances, and upholstered furniture, especially down the backs and sides. If you suspect an infestation, wrap the vacuum bag in a plastic bag and immediately set it in a tightly covered garbage container, preferably in the hot sun.

8. Be aware of little-noticed items like stuffed toys whose owners have long since grown up, mounted animal trophies, and insect collections (keratin pests are a real problem in natural history museums).

9. Vacuum and wash your pet's bedding frequently to prevent buildup of sheddings.

10. Watch your spices and cereals because these also draw carpet beetles.

11. If you have a piano, vacuum its felts frequently and keep a cake or crystals of true camphor under the cover.

12. If you're a violinist, vacuum the case often to protect your bowhairs.

13. Keep any hog- or sable-bristled art brushes in an airtight box.

14. Store fabric and yarn leftovers in tightly closed plastic bags.

15. When preparing clothes for out-of-season storage (see following page), treat the textile scraps at the same time.

Proper Handling and Storage

In winter, warm clothing and bedding are safe from fabric pests, if they're in regular use. Just moving the bedding around or wearing your woolens disturbs larvae, which drop off the textile before doing much harm. However, in summer, careful storage is essential.

Blankets, quilts, afghans, and throws should be laundered or dry-cleaned before storing. Dry cleaning destroys all stages of insect life but does not prevent reinfestation.

Sunning and Brushing

A long sunning followed by a brisk brushing, shaking, or beating is one of the most effective measures against fabric pests. Separate the clean items and hang them on a clothes line for at least 8 hours. Hang each suit part on its own hanger. As the sun moves around the clothes, any existing larvae drop off to get out of the bright light. While still outdoors, brush each garment vigorously; this crushes any eggs and sweeps any remaining larvae to the ground. Pay close attention to collars (both sides), pockets and flaps, pleats, and seams.

Note: Do not sun furs as this can cause fading.

No Clothesline? Use Your Freezer, Oven, or Pressing Iron

If you have no facilities for airing your clothes outside, you have other options. Putting small items—sweaters, scarves, woolen gloves—into your freezer for a few days will kill all stages of insect life. So will dry cleaning or pressing with a steam iron. Or, you can set your woolens in the oven for 1 hour at 140°F. Watch that plastic buttons don't start to melt; incidentally, chemical moth crystals can also damage buttons.

Storage Facilities Can Be Simple

You can use such common things as newspaper, adhesive tape, and cardboard boxes to store your fabrics.

- Put clean, insect-free sweaters, scarves, and blankets into heavy plastic bags or clean cardboard boxes. Seal these with adhesive tape.

- Brown wrapping paper or newspaper packages, sealed tight, also keep fabric pests out. If you use newspaper, first wrap the items in an old sheet or towel to prevent ink rub-off.
- Put large items in hanging zippered plastic bags, with their hanger holes securely taped. Crystals or cakes of real camphor increase the protection.
- The traditional cedar chest also provides safe storage. To be adequate, the chest must be airtight and made of at least 70 percent $\frac{3}{4}$-inch heartwood of red cedar. Renew its repellency every few years by brushing 100 percent cedar oil over the interior. Never paint the inside.
- If you don't have a cedar chest (or a cedar closet) you can gain some of cedar's benefits by hanging this wood's shavings (pet stores carry them) in a pillowcase in the closet with the cleaned garments.

Two additional points: Remember to sun, air, and brush valuable, rarely used items from time to time. And don't store vulnerable fabrics in the attic in the summer. That's the warmest part of your house and summer is when fabric pests are most active.

Treating Damaged Items

If this advice comes too late for some of your cherished woolens, you can have the damaged pieces that are worth repairing cleaned and the holes rewoven. It can be expensive but, properly done, the mend is invisible. Badly infested garments should be destroyed, either by burning or setting them in a plastic bag in a closed garbage container and baking the whole thing in the hot sun.

Herbal Repellents

Before the advent of chemical pesticides, many homemakers relied on herbs to hold off insect fabric-eaters. Garments so protected must be cleaned and stored as described in the previous pages.

Those legendary housekeepers, the Shakers, used bags of mint and tansy to protect their closets. Another old-fashioned barrier is a mixture of 2 handfuls each of dried lavender and rosemary with 1 tablespoon of crushed cloves and small pieces of dried lemon peel, all placed in gauze bags to be kept in dresser drawers.

The *Organic Farmer* recommends any of these as moth repellents:

Rosemary   Citrus peel
Southernwood Cloves
Lavender   Spearmint
Cedar   Tansy
Sassafras   Santolina
Bay leaves

Note: Many of these substances attract pantry pests, so you could be exchanging one insect problem for another.

How to Insectproof a Storage Closet

Choose a seldom-used closet.
Fill wall and ceiling cracks with putty or plastic wood.
Weatherstrip the door with synthetic weatherstripping, or
Seal cracks around the door with tape.

How to Insectproof a Trunk or Storage Chest

Seal any holes or cracks.
Tape the lid if it does not fit tightly, or
Wrap the whole chest in heavy paper and tape it shut.

SOURCE: Adapted from *Carpet Beetles and Clothes Moths,* Berkeley, CA: Division of Agricultural Sciences, University of California, rev. August 1979. Leaflet 2524.

A Carpet Beetle Trap

In the 1930s, pest expert Hugo Hartnack devised an ingenious trap for carpet beetle larvae: Lay some American cheese on soiled woolens or furs in the corners of an infested room. This will attract the crawlers, which you can then easily catch and drown in hot water.

Nature of the Silverfish

Silverfish can get into your home via several routes.

1. You may carry them in in secondhand books, cardboard boxes, or old papers.
2. If your home is new, they may have come with the wallboard and green lumber, and found a feast of wood shavings, sawdust, and wallpaper paste. Humidity from newly poured cement gives them the dampness they need.
3. These long-lived nocturnal insects may be making their way inside from flowers near your home's foundation.

Once indoors, they hide and lay their eggs in wall cracks and behind baseboards. Only their scaly, silvery appearance gives them their name for they do not swim. (See Figure 10.4.) If you find one in the tub or sink, it has fallen in accidentally and can't climb up the slippery sides.

Apart from the insect itself, the first signs of silverfish are damaged papers or books. They eat the sizing on paper and chew ragged holes in the paper itself. They also eat cereals, dead insects, linen, cotton, silk, paste, and glue (as used in bookbindings). They can be found in any room of the house.

Shut Silverfish Out

Preventing and controlling silverfish is much less traumatic than confronting their ravages on irreplaceable books and documents. Prevention begins outside.

1. Close off any holes around pipes entering the walls.
2. Relocate flowers growing around foundations some distance from the structure because these insects like the mulch in flower beds.
3. If you collect old books, check all new acquisitions carefully before setting them on your shelves.
4. Clean out bookcases periodically, shaking out the books.
5. Repair any plumbing leaks.
6. Check any lined draperies you have; silverfish often hide between the lining and outer fabric.

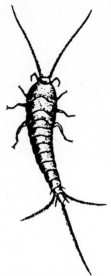

Figure 10.4. *Insect and Rodent Control: Repairs and Utilities.* War Department Technical Manual, TM-5-632, October 1945.

Wipe Silverfish Out

Thorough cockroach control invariably eliminates silverfish. Like cockroaches, these insects are poisoned by technical boric acid. The powder should be blown with a bulb duster into any baseboard crevices and around door and window frames. Control may take several weeks.

A Trap for Silverfish

A small glass jar, its outside covered with masking or adhesive tape, makes a good silverfish trap. The insect crawls up the tape, falls into the jar and can't climb up the slippery inside walls. Some authorities advise baiting the jar with a bit of wheat flour; others say this isn't necessary. Set the trap in a corner for greatest effectiveness.

With all these small marauders waiting to savage our clothes, home furnishings, and records, it's a wonder we all aren't threadbare. The truth is that reasonably careful housekeeping holds their damage to levels we can usually tolerate, without chemical insecticides.

CHAPTER 11

Hidden Vandals: Termites

In the seventeenth century, La Rochelle, the Huguenot port from which Champlain and Cartier sailed for the New World, fell to the cannons of Cardinal Richelieu. Rebuilt, the city almost fell again 2 centuries later—this time to an insect as destructive as any cannon. When tropical termites invaded the town in the nineteenth century, whole streets were undermined. The arsenal and chief government building had to be shored up, and all the town records were chewed to a pulp.

During World War II, crates containing military supplies, temporarily stacked on the ground in New Guinea, disintegrated when they were moved a few weeks later, because they had been invaded by subterranean termites.

First Cousin to the Cockroach

Like their cockroach ancestors, termites speed the transformation of dead forest trees into soil nutrients, a process that might otherwise take hundreds of years. In earlier times these "white ants" eased the pioneers' heavy task of settling and cultivating new land. After trees were felled, the insects gradually cleared away the stumps, in the process enriching the soil for the farming that followed.

In Sri Lanka three-fourths of the clay-like land would be unfit for

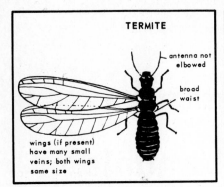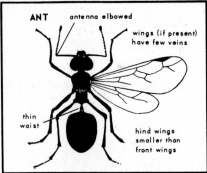

Figure 11.1 Major differences between winged ants and termites. (*Termites and Other Wood Infesting Insects*. Division of Agricultural Sciences, University of California, Leaflet 2532, 1981.)

agriculture without the teeming termites loosening the soil so rain can penetrate it. Tropical termite mounds, often 20 or 30 feet high, are made of material so tough that nearby humans use an abandoned hill as a baking oven or metal smelter. Crushed, the substance makes a resilient, long-lasting surface for a tennis court.

Termite Damage

Despite its benefits, this insect becomes an expensive pest where human beings use wood to make their homes. In the United States alone about 330,000 homes are treated every year for termites. Authorities claim that the annual wreckage equals that caused by fire.

Unless you live in a warm region, you probably have never seen a termite. If you did, you may have confused it with a winged ant. (See Figure 11.1.)

Termite Species

Scientists have divided termites into two major groups: those that live in the earth and those that live above it. Subterranean species cause the most damage.

Buildings in warm, moist regions are particularly vulnerable to subterranean termites. Along the Gulf and Pacific coasts, the drywood termite, which needs no soil, causes considerable havoc. The only state free of termites is Alaska. Figure 11.2 maps the areas of relative termite activity in the United States.

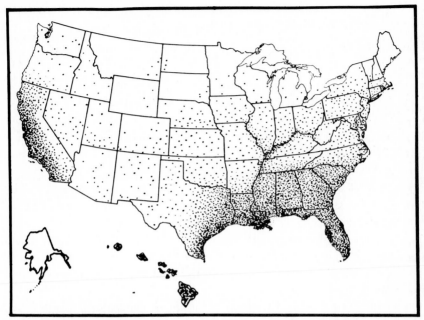

Figure 11.2. Relative hazard of termite infestation in the United States. [*Subterranean Termites, Their Prevention and Control in Buildings*. Washington, D.C.: United States Department of Agriculture Home and Garden Bulletin 64, 1979 (adapted.)]

Within the last 20 years another subterranean termite, the Formosan, has crossed the ocean to Texas and Florida. More aggressive in its tunneling than our native troublemakers, the Formosan termite is also more resistant to soil poisons that control other species. New, nonchemical techniques that infect other subterranean termites with a fatal disease are ineffective against the Formosan. The species bite off the legs of a sick termite so it can't spread the contagion throughout the colony.

A Worsening Problem

Even without the immigrant from Taiwan, the termite problem seems to be getting worse. Patios, breezeways, attached garages, all increasingly popular, open the way for termite invasions. Slab construction, now widespread, makes control difficult. As our virgin forests, with their mature trees, disappear, more homes are built with sapwood, which is more susceptible to termites. Finally, central heating encourages the insect's activity year-round.

Nature of the Termite

Subterranean termites live in colonies, which can number as many as a quarter of a million individuals; they nest in the ground but feed above it. To reach wood, they tunnel up, building flattened shelter tubes to protect themselves from predators where they must emerge from the soil. When not feeding they return underground to replenish their body moisture.

Drywood termites nest in the wood they eat and never need ground contact. Their colonies are relatively small. Because they can go without moisture for long stretches, they're often carried in furniture and crates from one area to another. Not surprisingly, drywood infestations spread easily.

Because these insects don't like to be shaken, objects such as railroad ties and musical instruments that vibrate frequently are relatively safe from assault. If you need a good reason to play your piano, this is it.

When Is a House in Danger from Termites?

Termites are an ever-present danger to property owners, even those with brick homes. A crack in mortar or cement only $\frac{1}{32}$-inch wide lets them crawl under the brick to the wood. One species has been found on the fifth floor of a concrete fireproof building in San Francisco.

Where subterranean termites are not so abundant, a house that meets modern building codes could remain sound for many years. On the other hand, if it were built over or near an existing colony, winged reproductives could show up after only 3 or 4 years. Structural damage might be evident some years after that. Along the Gulf Coast some of the ground is continuously infested; a house can be attacked as soon as it's built. Drywood termites, common along the west coast and the Caribbean, can also invade immediately.

Homes over 35 years old are more likely to have termite damage. In some regions every home on a block in an older neighborhood may show signs of termite invasions.

Signs of Termite Infestation

Termites do their dirty work out of sight, deep within our walls or furniture. By the time some of their destruction is visible, they are well established and are difficult to eradicate. However, they send out some early signs that can warn you of a developing problem.

1. **Swarming of the reproductives,** the potential "queens" and their consorts, as they fly out of the colony is one sure clue to all termite species. (Ants also have reproductive swarms.) After a brief mating flight, the sexually mature insects shed their wings (so do ants) before burrowing into wood to begin laying eggs. A mass of tiny wings scattered on the ground tells you that termites (or ants) are thriving in the area.

2. **Shelter tubes,** built of particles of earth and wood cemented together with a gluelike secretion, are a sign of subterranean termites. The tubes may be clinging to a foundation or hanging from a joist or girder.

3. **Wood flooring with dark or blistered areas** easily crushed with a screwdriver or kitchen knife signify subterranean termites. To verify your suspicions, rap on the suspect wood. If you hear an answering tapping, rather like the ticking of a watch, the subterranean soldiers are alerting the colony to danger by banging their very hard heads against the gallery walls.

4. **Fecal pellets** are the calling card of drywood species. Roughly oval and six-sided, the pellets are generally seen near the pests' "kick-out" holes. Termites are excellent housekeepers, sweeping all waste out of their galleries.

Wipe Them Out

When You Think You Have Termites

Although this book focuses on controlling pests without chemicals, termite control is one area where chemicals are almost always necessary. Applying them is a job for a well-trained professional. Homes have been made unlivable and families have become ill because of poor termite-control work.

How do you go about finding a competent pest control company? Very carefully.

Mishandling of Chlordane a Serious Problem

Chlordane, probably the most widely used poison for subterranean termites, is strongly suspected to cause cancer. It's also linked to other serious illnesses. It is very long-lasting (a big plus in termite control) and accumulates in fatty tissues. Chlordane is not for the inept or the amateur.

In Los Angeles, a beautiful older home, treated for subterranean termites in the late 1970s, became a poisonous trap for its owners. For a year after chlordane was sprayed into the structure's crawl space, family members suffered repeated headaches and nausea. In an effort to make the house livable, the owners had the entire heating and ventilating system reworked, but no one—not the EPA, the county health authorities, nor the National Academy of Sciences—could define for them a safe level of exposure to chlordane.

In 1983, a young Long Island family had to see their contaminated home bulldozed because of incompetent termite treatment.

These are extreme cases, but the EPA thinks as many as 80 million homes in the United States may be contaminated with chlordane residues. The mishandling of this termiticide is a serious problem. The chemical's only legal use is for killing termites; it's a restricted-use substance. However, it is not necessarily being applied by someone certified by the state or federal government to do so. Applicators don't always follow the directions on the chemical's label.

Dursban is another hazardous substance widely used against termites. Its long-term effects are weaker than chlordane but in the short run it's more toxic.

Methods of Treatment for Subterranean Termites

Treating techniques for underground termites depends on the structure's foundation. A slab may require drilling close to the infestation and injecting a termiticide into the soil beneath the house. In another approach, insecticide-carrying pipes are introduced horizontally across the slab to saturate the soil. A house with a crawl space is the easiest to treat. Applicators dig a trench around the foundation, pour the chemical into the trench, and replace the soil.

In a nonchemical approach, Spear, a promising biological substance in use for several years, is introduced into the soil. Its microorganisms home in on the pests, which sicken and die. Other termites in the colony get the disease when they consume the victim (termites are cannibals).

Treating for Drywood Termites

Drywood termites must be handled very differently from those invading from the soil. If the infestation is clearly localized, "drill and treat" is the most common method. Holes are drilled to the insects' galleries and chemical poisons injected into them. A new, nonchemical treatment, Electrogun, sends a current of electricity through the galleries, electrocuting the pests or so disrupting their life systems that they soon die. It has been used for several years with good results. It is very safe.

Because it's difficult to determine the extent of a drywood infestation, pest control operators prefer to use fumigation against these pests. After drill-and-treat the question always remains, "Did they miss some?" After fumigation the answer is an unqualified "No."

If the termites are only in a piece of furniture you can have it fumigated away from your home. Some pest control companies and some county agricultural offices provide this service. The National Academy of Sciences says that exposing infested furniture to subfreezing temperatures destroys the termite colony. Unfortunately, if you live where the mean annual temperature is above 50 degrees, your home is more vulnerable to termites and you may have to wait a long time for a freeze.

Who Controls the Pest Controllers?

In the United States the federal government sets the nation's standards for pest control and the individual states use them to certify pest control operators as competent. State tests cover such subjects as:

- Thorough understanding of chemical labels.
- Safety, including precautions in treated areas; protective clothing and gear; poisoning symptoms and first aid; proper handling, storage, and disposal of pesticides.
- Recognition and biology of pests.
- Thorough understanding of pesticides.
- Proper use and maintenance of equipment.
- Proper application techniques; preventing drift and pesticide loss to the environment.

The state test, however, doesn't mean that the people who come to treat your property are experts in all these areas. In some states the applicator may not have taken any test at all and *may not even know how to read*. A pest control company can legally operate with only one certified staff member who oversees those actually applying the chemicals.

Federal law requires that where a certified applicator will not be on the scene, he or she must provide detailed guidance to the actual workers. Provision must also be made for contacting the supervisor quickly.

How to Find a Competent Pest Control Company

Supervision is the key to safe, effective commercial pest control. How do you find the right company for the job?

1. Get a recommendation from a friend who's had a termite problem and been satisfied with the company that handled it.

2. Check the yellow pages of the telephone directory. The ads there, ranging from the catchy to the business-like, may present a confusing choice. My own preference is to call companies that avoid the bug cartoons and solicit my business in a straightforward way.

3. Contact at least three. It may or may not be to your advantage to talk only with companies that have been around a long time. The outfit that contaminated the Los Angeles home mentioned earlier had been in business for 50 years.

Ask the Right Questions

Begin to ask questions on the telephone. If your state supervises pest control closely—California, Maryland, Florida, and North Carolina are among the best—your questions may not need to be so detailed as these.

1. Is the salesperson who'll come to the house licensed to make recommendations on termite control? If the answer is no, call another company.

2. How much of the firm's business is in termite control? In areas where termites abound, they make up about 30 percent of the average company's business, if they're licensed for this category.

3. How will the applicators be supervised?

4. Are the applicators literate?

5. How are they trained? Do they and their supervisors receive continuing education in pest control?

6. What insurance does the company carry? General liability and worker's compensation should be the minimum. Coverage for errors and omissions would also make sure the company could reimburse you if the crew fouls up.

7. Are they familiar with and do they use advanced, nonchemical controls such as Spear and Electrogun?

Questions about Fumigation

Fumigation, the usual treatment for dispersed infestations of drywood termites, requires a different set of questions. Safety must be the absolute prime consideration in fumigation. The gases used are odorless, colorless, tasteless, and deadly. The company's representative should be crystal clear about how you and your property will be protected.

1. How will foods be protected?

2. What should be done with gas-sensitive plants both inside and outside the house?

3. Will photographic equipment be affected?

4. What will happen to valuable art works like oil paintings?

5. Will rubber and leather items be affected? If the fumigant is methyl bromide, then rubber, including that in your carpet padding, will develop a persistent, unpleasant odor. Leather may also be affected. Sulfuryl fluoride affects neither. Some metals may corrode.

6. Will weather affect the fumigation? The colder the day, the longer the gas takes to kill the pests and decompose. If the temperature is expected to drop below 50°F, the pest control operator may heat the fumigant.

7. How will my vacant home be secured? Strong warning signs with skull and crossbones will alert honest people if the tarpaulins don't, but supplemental locks should be installed on all doors. You'll probably have to surrender your keys. Even so you run the risk of burglary so put your valuables in a bank vault or other safe place.

Richard Yaussi, a pest control operator in Long Beach, California, had two fumigation jobs burglarized in 1 month. The fumigants are laced with tear gas to make them detectable, but the thieves, using roll after roll of paper towels and toilet paper to wipe their streaming eyes and noses, took what they wanted.

8. How will I know if my house is safe to reoccupy? Wearing protective gear (probably gas masks), applicators open the doors and windows and turn on fans. A special device determines whether or not any gas is left. There should be none.

9. How long will the fumigant protect my home? As soon as the gas has dissipated, your house can be reinfested immediately. There's no residual effect. Discuss with the pest control operator the desirability of dusting wall voids and attic with a long-lasting sorptive dust such as Dri-die to kill any future pests thinking of setting up housekeeping.

10. How much does fumigation cost? Fumigation is expensive. A friend of mine had his average-size, two-story home fumigated in 1984. The bill was $1,100, not counting his motel for two nights, restaurant meals, and board for his bassett hound.

Check Further

Be wary of a company that tries to high-pressure you. Even if you do have termites, your house is not going to tumble down tomorrow or even next week. Houses have stood, infested, for 50 years.

Besides contacting two or three customers of each company, call the state agency that oversees the industry and ask how many violations each has had filed against it in the past year or two. The agency may be part of the health or agriculture department. It may be called the Structural Pest Control Board, as in California; the Office of Paints and Chemicals, as in Virginia; or something else.

Probably no company has a spotless record. Pest control work is complicated, difficult, and often dangerous. Human beings, no matter how intelligent and conscientious, make mistakes. Common sense says to choose the concern with the fewest marks against it. Take their relative size into account, for a small concern could have fewer violations than a large one, yet do no better a job.

A company may have a high percentage of error-free applications but may be unpleasant if there's a misunderstanding about your contract. To judge the kind of people you'll be dealing with, check with your community's consumer affairs bureau, chamber of commerce, or Better Business Bureau.

When the applicators arrive at your home, they, their clothing, their truck and equipment should be clean, says Robert LaVoire, president of a large Los Angeles pest control company. And there should be no odor of liquor about them. Termite control demands a clear head.

Shut Them Out

Knowing something about how to prevent termites will lessen your chances of needing professional control. Termite prevention begins with sound construction.

Preventing Subterranean Termites

Standard building codes carefully followed are the best protection against subterranean species. If your home is under construction, be aware of the following steps to forestall the problem.

Make the Structures as Dry as Possible

1. The lot should be graded to let water drain rapidly away from the house and its adjacent structures.
2. Gutters, downspouts, and foundation drains should all guide water clear of the building.
3. Doors, windows, roof valleys, and chimneys need to be adequately flashed.
4. Attics and crawl spaces should be well vented to prevent accumulations of moisture.
5. Walls should be fitted with vapor barriers.
6. In very moist regions, the soil under a slab-based house should be shielded with a heavy polyethylene sheet before concrete is poured.

Soil Poisoning

Slab construction, once thought to be effective termite proofing, may actually open the way for an infestation. Temperature changes, earthquakes, sonic booms, natural earth shifts all can crack concrete, giving termites easy access. One of the surest defenses is soil poisons applied before building begins. It's a relatively safe procedure, says the United

States Department of Agriculture. Tests have shown that oil-soluble insecticides stay in place, moving neither sideways nor down toward the
water table, and can protect a structure for 25 years.

A Poured Foundation

Poured concrete foundations, easily inspected, are the most termite-
resistant form of residential construction. In addition, a reinforced concrete cap 4 inches high between foundation and subflooring will expose
any termite tubes. Metal termite shields, once thought more effective than
the concrete cap, are difficult to install properly and so in actual practice
don't work well.

Proper Handling of Wood

Since termites must eat wood or wood products like paper, wood pulp,
and fiberboard, the critical factors in building are:

1. Treating the wood with repellents or insecticides before it's incor-
    porated into the structure.
2. Permitting no wood to touch soil at any point.

The Uniform Building Code, followed in about twenty states, requires
the use of treated wood wherever it's in contact with masonry or concrete. Since subterranean termites have to tube across treated members to
reach the untreated ones, an infestation is easily detected.

Standard codes also require that:

- No wood penetrate concrete to the soil below.
- There be adequate clearance between structural wood and the soil.
- A crawl space large enough for thorough termite inspection be
  provided.

Most important in home building is meticulously clearing away all
wood debris when the house is finished, especially from earth filling
porches, steps, and patios.

Poorly handled wood scraps are responsible for over half the infestations of subterranean termites. Yet, "I've never heard of a building inspector digging through backfill to look for wood debris," says Dr.
George Rambo, Director of Technical Operations for the National Pest
Control Association. If your home is under construction, make sure that

all backfill is free of wood debris, even if you have to take a shovel and dig through it yourself.

Figure 11.3 shows twenty points of careless construction and home maintenance that can open the way for subterranean termites.

Preventing Drywood Termites during Construction

If yours is an area with drywood termites, before the house is decorated, consider having a pest control operator drill holes in the plaster or wallboard and blow in a silica aerogel dust. This will remain in place indefinitely, killing any insect that crawls over it. You should also arrange to have an annual termite inspection.

Incidentally, houses with attached garages whose doors are kept open much of the time, have a greater chance of invasion by drywood termites, which are blown in from dead trees and telephone poles.

Inspection for Termites

A professional termite inspection is your best insurance against infestation in a house already built. Such an inspection is a difficult, time-consuming job requiring agility and expertise. The inspector must probe all your home's hidden recesses, crawling through dark, cramped places to examine drains, floor furnaces, and attics. Be prepared to pay a reasonable fee for the service.

You should receive a formal, written report of the findings, including a clearly labeled diagram of the house indicating any infestations, all termite-favoring conditions, and areas that could not be inspected. If you feel the inspection isn't thorough enough, arrange for a second. Figures 11.4 and 11.5 are a standard structural pest control report form and a sample diagram of the inspected house.

Your state may also require a second report to be filed with the agency overseeing the pest control industry. This report indicates what remedial work has and has not been completed. In California, for a nominal fee, you can get from the Structural Pest Control Board a copy of every report filed within the past 2 years for any property you're thinking of buying. It would be prudent to track down such reports no matter what state you live in.

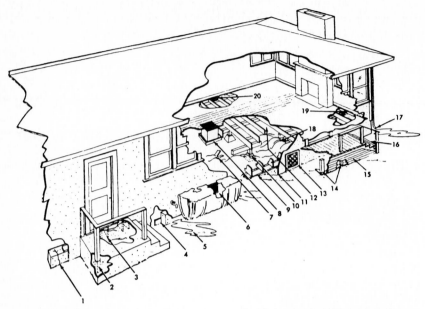

1. Cracks in foundation permit hidden points of entry from soil to sill.
2. Posts through concrete in contact with substructural soil. Watch door frames and intermediate supporting posts.
3. Wood framing members in contact with earthfill under concrete slab.
4. Form boards left in place contribute to termite food supply.
5. Leaking pipes and dripping faucets sustain soil moisture. Excess irrigation has same effect.
6. Shrubbery blocking air flow through vents.
7. Debris supports termite colony until large population attacks superstructure.
8. Heating unit accelerates termite development by maintaining warmth of colony on a year-round basis.
9. Foundation wall too low permits wood to contact soil. Adding top soil often builds exterior grade up to sill level.
10. Footing too low or soil thrown against it causes wood-soil contact. There should be 8 inches of clean concrete between soil and pier block.
11. Stucco carried down over concrete foundation permits hidden entrance between stucco and foundation if bond fails.
12. Insufficient clearance for inspection also permits easy construction of termite shelter tubes from soil to wood.
13. Wood framing of crawl hole forming woodsoil contact.
14. Mud sill and/or posts in contact with soil.
15. Wood siding and skirting form soil contact. Should be a minimum of 3 inches clearance between skirting and soil.
16. Porch steps in contact with soil. Also watch for ladders and other wooden appurtenances.
17. Downspouts should carry water away from building.
18. Improper maintenance of soil piled against pier footing. Also makes careful inspection impossible.
19. Wood girder entering recess and foundation wall should have 1 inch free air space on both sides and end and be protected with a moisture impervious seal.
20. Vents placed between joists tunnel air through space without providing good substructural aeration. Vents placed in foundation wall give better air circulation.

Figure 11.3. Faulty home construction and maintenance promoting subterranean termite infestation. (Walter Ebeling. *Urban Entomology*. Berkeley, CA: Division of Agricultural Sciences, University of California, 1978. (rev.) After Eleventh Naval District, San Diego.)

STANDARD STRUCTURAL PEST CONTROL INSPECTION REPORT
(WOOD-DESTROYING PESTS OR ORGANISMS)
This is an inspection report only - not a Notice of Completion.

ADDRESS OF PROPERTY INSPECTED	BLDG. NO.	STREET	CITY	DATE OF INSPECTION
			CO. CODE	

FIRM NAME AND ADDRESS

Affix stamp here on Board copy only
↓ A LICENSED PEST CONTROL ↓
OPERATOR IS AN EXPERT IN
HIS FIELD. ANY QUESTIONS
RELATIVE TO THIS REPORT
SHOULD BE REFERRED TO HIM.

FIRM LICENSE NO.	CO. REPORT NO. (if any)	STAMP NO.

Inspection Ordered by (Name and Address)_____

Report Sent to (Name and Address)_____

Owner's Name and Address_____

Name and Address of a Party in Interest_____

INSPECTED BY: LICENSE NO.   Original Report ☐   Supplemental Report ☐   Number of Pages

YES	CODE	SEE DIAGRAM BELOW	YES	CODE	SEE DIAGRAM BELOW	YES	CODE	SEE DIAGRAM BELOW	YES	CODE	SEE DIAGRAM BELOW
		S-Subterranean Termites			B-Beetles-Other Wood Pests			Z-Dampwood Termites			EM-Excessive Moisture Condition
		K-Dry-Wood Termites			FG-Faulty Grade Levels			SL-Shower Leaks			IA-Inaccessible Areas
		F-Fungus or Dry Rot			EC-Earth-wood Contacts			CD-Cellulose Debris			FI-Further Inspection Recom.

1. SUBSTRUCTURE AREA (soil conditions, accessibility, etc.)_____
2. Was Stall Shower water tested?   Did floor coverings indicate leaks?_____
3. FOUNDATIONS (Type, Relation to Grade, etc.)_____
4. PORCHES . . . STEPS . . . PATIOS_____
5. VENTILATION (Amount, Relation to Grade, etc.)_____
6. ABUTMENTS . . . Stucco walls, columns, arches, etc._____
7. ATTIC SPACES (accessibility, insulation, etc.)_____
8. GARAGES (Type, accessibility, etc.)_____
9. OTHER_____

DIAGRAM AND EXPLANATION OF FINDINGS (This report is limited to structure or structures shown on diagram.)

General Description_____

Signature_____

YOU ARE ENTITLED TO OBTAIN COPIES OF ALL REPORTS AND COMPLETION NOTICES ON THIS PROPERTY FILED WITH THE BOARD DURING THE PRECEDING TWO YEARS UPON PAYMENT OF A $2.00 SEARCH FEE TO STRUCTURAL PEST CONTROL BOARD, 1430 HOWE AVENUE, SACRAMENTO, CA. 95825.

Figure 11.4. Standard structural pest control report. (*So, You've Just Had a Structural Pest Control Inspection.* Division of Agricultural Sciences, University of California, Leaflet 2999, 1980.)

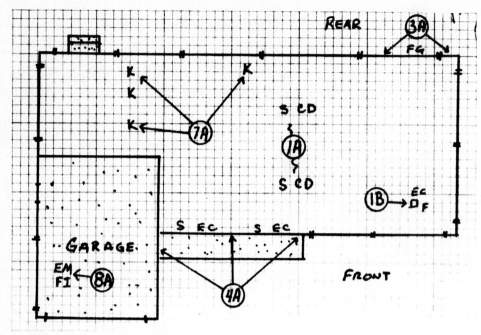

Figure 11.5. Termite inspector's diagram of a house. (*So, You've Just Had a Structural Pest Control Inspection*. Division of Agricultural Sciences, University of California, Leaflet 2999, 1980.)

Checklist for Keeping Your Home Termite-Free

✔ All plumbing should be kept in good repair.

✔ Shower pan should be free of leaks.

✔ Basement air vents should be fully exposed, not overgrown with shrubbery.

✔ When watering, avoid sprinkling stucco or wood siding.

✔ Fill any cracks in masonry or concrete with cement grout, roofing grade coal-tar pitch, or rubberoid bituminous sealers.

✔ Keep gutters and downspouts in good repair.

If ants are hard at work around your foundation, leave them alone—
and be grateful. They are nature's termite inspectors, always on the job.
Termites have many predators—birds, lizards, beetles—but their greatest
foe is the ant. War between the two is eternal.

CHAPTER 12

Garden Insects— Part of the Landscape

A homeowner is showing his rose garden to a friend, who's admiring the vigorous, fragrant blooms.

"Mine don't look anything like these," says the friend, "What systemic do you use?"

The garden's owner raises his thumb and forefinger. "These," he replies, and rubs them gently around a stem, crushing one or two aphids. "And this," he adds, pointing to a lady beetle foraging on a leaf.

His flourishing shrubs have never been sprayed with anything more lethal than a dormant oil spray. Around the entire bed, graceful onion and garlic plants stand guard. The roses get the simplest care. They're never watered after noon; and during their growing season they receive a monthly ration of rose food.

Integrated Pest Management in Your Garden

This garden is a fine example of integrated pest management (IPM). Instead of insecticides, the gardener uses a number of "compatible means to obtain the best control with the least disruption of the environment," as ecologists Mary Louise Flint and Robert van den Bosch have described it. There are many ways you too can apply this new/old technology in your garden to control some of the most common plant pests.

Some IPM strategies include:

- Cultural controls like interplanting and crop rotation
- Companion and trap plants
- Homemade sprays
- Encouraging natural predators
- Mechanical barriers and traps
- Repellents

Organic plant protection is a vast subject, beyond the scope of this book to deal with in any detail. However, the suggestions in this chapter are those experienced horticulturists have found to work against the pests most likely to be feeding in your garden. These measures should start you thinking about devising some tactics of your own to keep the hungry hordes under control without fouling the air you breathe and the water you drink.

Cultural Controls

The reason U.S. farmers rely heavily on chemical pesticides is that they plant broad spreads of a single crop, which build up a great potential for that plant's pests. As a home gardener you don't have to do that.

Interplanting

By setting out your plants somewhat at random instead of in tidy rows, you will make it harder for pests to find a continuous food supply. Chinese farmers employ this method, planting some wheat, with some potatoes nearby, and maybe some rice so that insects that feed on any one of these crops are confined to a small area.

Crop Rotation

If you plant tomatoes in one spot for several years, tomato pests have a chance to build up their populations. You also deplete the soil of nutrients so you have more pests attacking weaker plants. To avoid this, plant a different crop in the same location each year, such as tomatoes one year, squash the next, and peas the following year. Or consider letting your flower and vegetable plots trade places.

In choosing vegetables, try to avoid pest-prone plants like cucumbers, melons, radishes, and broccoli. Instead, use such crops as beans, peas, and spinach—all hardier species.

It's a good idea to put in more vegetables than you really need so that you can tolerate some loss to insects or plant disease. You can also time your plantings to avoid a particular crop's pests. Your county farm agent can advise you on this, as well as on proper watering and fertilizing practices.

Once you've brought in your harvest, pull up all the plants that are past their growing season so that their pests have no place to overwinter and will die. In the fall turn the soil over to let the birds deal with grubs and maggots. Those the birds miss will die of starvation or cold or simply dry up.

Consider letting a small stand of weeds in an out-of-the way corner stay for the winter. It will serve as harborage for a few pests, enough to keep the predators around for the next season.

Companion Planting

Cultural controls are like housekeeping transferred out of doors. The same strategies of rotation, plenty of fresh air, and cleanliness work as well outside as they do inside. These methods, says the National Academy of Sciences, "are often economical and dependable, although seldom spectacular." In some parts of the world farmers use them as a matter of course. They also use a strategy called companion planting.

In Russia, farm workers set mustard plants among their cabbages to increase the effectiveness of parasites that live on cabbage worms. In Hawaii, growers surround squash and melon fields with a few rows of corn which is very attractive to the melon fly, a major pest. Onions and garlic growing here and there throughout your garden will repel insects. Allyl, their active ingredient, has been found by University of California scientists to be an effective mosquito repellent. Strong-smelling herbs like rosemary, peppermint, pennyroyal, southernwood, wormwood, lavender, and sage planted among crops are also said to hold off insect pests.

Handpicking

Although too time-consuming to be practical on a commercial farm, handpicking of insects like hornworms on tomatoes or bagworms on ornamentals is feasible for the home gardener. You can either gather the pests

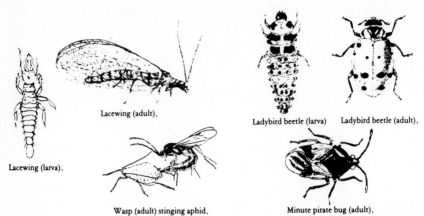

Figure 12.1. Beneficial garden insects. (*Controlling Insects, Diseases, and Related Problems in the Home Vegetable Garden.* Division of Agricultural Sciences, University of California, Leaflet 21086, 1979).

in a plastic bag and dispose of them with the garbage or put them in a jar with a little water, where they will decompose. When the same species appears again, set the open jar under the plant they're invading. This can be a strong repellent because most animals tend to avoid their own waste.

Biological Controls: Predators and Insect Diseases

In holding down the garden pest population you have some formidable allies whose services cost you nothing and who never pollute the environment. They don't have to be sprayed or dusted on and they often go to work long before you're even aware that you have a pest problem. These are the predators—birds and insects, lizards and toads—that feed on the pests that feed on your plants.

Insect predators are the world's most thorough insecticides. Many pest young—eggs, larvae, and pupae—are immune to chemicals but a predator can swiftly lay its egg on the young of another insect and the newly hatched predator consumes its hapless victim long before it has a chance to do any damage. And the best part is that the predators cost you nothing. Figure 12.1 shows a few of the most helpful garden insects.

Insecticides often kill off beneficial insects and, though they may bring immediate relief, in the long run there's serious risk of an upsurge of a secondary pest that was of no importance before but now, free of its enemies, multiplies astronomically.

California offers a classic example of successful biological control disrupted by the widespread use of organic chemicals. In the late nineteenth century a scale insect was destroying the state's young citrus groves. The vedalia beetle, a kind of lady beetle, was imported from Australia, where it preyed on scale insects. The beetle flourished in California and within 2 years had almost completely routed the scale insects. Then in the late 1940s DDT was introduced as the ultimate insecticide, which California citrus growers adopted as enthusiastically as other farmers did. Along with various pest species, the vedalia beetle was also eradicated, and the scale insect rebounded with devastating effect. The damage to the groves was the worst it had been in 50 years.

Community Cooperation

One problem for the home gardener practicing biological control is that while you're wooing the predators, pesticides may be drifting over from the neighbors' yards. If you can enlist the community's cooperation to hold off on the chemicals and give the predators a change to do their work, you'll all enjoy a more healthful environment.

Commercially Supplied Predators

Natural predators like lady beetles, trichogramma wasps, and lacewings can be bought from commercial insectaries for release on your property. Reviews on this are mixed. The immigrants don't always take to their new surroundings and may not give you the hoped-for results. Mass releases of commercially bred predators have been successful in Great Britain when aimed at aphids and spider mites in greenhouses and, as we've seen, in the United States against citrus pests. However, like other aspects of pest control, nothing is guaranteed.

Some Useful Insect Predators

Among the most helpful insect predators—all found in urban areas—are lady beetles, dragonflies, lacewings, hover flies, ground beetles, centipedes, ichneumon flies, and miniwasps. Larger predators include snakes, toads, lizards, hawks, owls, ravens, seagulls, weasels, bats, raccoons, badgers, and opossums. Some would add praying mantids. However, they really aren't that beneficial since they eat everything in sight, including each other and truly helpful insects.

Making the Pests Sick

Organisms causing insect diseases are also potent pest controllers. The most common, *bacillus thuringiensis,* is available as an insecticide in garden shops. It poses no threat to mammals and usually cause no problems with helpful insects.

Another disease agent, milky spore, was a major factor in bringing the Japanese beetle to heel in the 1940s and 1950s. It is still used.

Homemade Plant Sprays

Organic gardeners have long known that homemade sprays can be of great help in protecting plants. The simplest is clear water. Often a strong syringing from a garden hose is enough to dislodge plant pests.

Soap sprays have been used successfully since early in the nineteenth century, when fish- and whale-oil soaps were the rule. Modern soaps or biodegradable detergents at a concentration of 1 or 2 percent (roughly 1 or 2 tablespoons to 1 gallon of water) also work.

In tests conducted by the California Department of Agriculture, liquid soaps and detergents have good control, in some cases comparable to that of synthetic organic pesticides. Mites, aphids, psyllids, and thrips were all killed by the sprays, which seem to work in two ways: by dislodging the pests and by smothering them. Ivory liquid detergent gave the most consistent control. The scientists suggest that you rinse off the plants with clear water several hours after the soap spraying to reduce leaf burning.

Mechanical Barriers against Tree Pests

A number of mechanical barriers can keep crawling pests out of shade and fruit trees.

- The National Academy of Sciences recommends wrapping sticky bands around the trunk 2 to 4 feet from the ground to block cankerworms, gypsy moths, cicadas, and ants who pasture their aphids in trees.

- The United States Department of Agriculture says that flypaper applied to a band of heavy paper 6 to 8 feet wide and then tacked around the trunk also works. Fill the rough places in the bark with cotton batting so pests can't crawl under the paper.

- Another barrier is cotton batting several inches wide wrapped tightly around the trunk. Tie the batting tightly with string near the bottom edge and the upper portion of the band. Turn the upper portion down over the lower to trap crawling pests.

- Mosquito netting, 16 wires to the inch and 14 inches wide, can also guard a tree. Tack it to the trunk so that it fits tightly at the top and flares out half an inch or more at the bottom. Smooth or fill all rough places on the bark so caterpillars can't crawl under the top edge of the band.

Controls for Specific Pests

The following nonchemical controls have been found to work for some of the more common garden pests. You may have to try several before you achieve the results you want.

Aphids

Most plants can sustain a moderate number of aphids without harm, so finding a few on your favorite roses is no reason to bring up the chemical howitzers. Steps you can take to minimize these tiny, pear-shaped pests include application of sticky bands, soap sprays, rubbing, and pruning.

Ants fight off aphid predators so keeping ants away from their "cattle" gives the predators a free hand. Sticky bands wrapped around plant stems will do just that. Aphids are food for a number of insects and their young, including the well-known lady beetle, lacewings, syrphid flies, daddy longlegs spiders, and praying mantids. Birds also find them tasty. Although aphids' reproductive rate is awesome, predators and natural forces like cold rains, temperature extremes, fungus, and bacterial disease hold their numbers down.

You can help keep their population within tolerable limits. Soapy water, as mentioned, sprayed at a firm pressure washes aphids off plants or smothers them. Water alone can also do it, but a little soap increases its effectiveness. Or you can rub infested plant parts gently between your thumb and forefinger, or wipe them with a damp cloth. Pruning infested plant parts also helps.

To control aphids on trees, thin out the dense inner canopy that provides the pest with protected living quarters. When forced to the tree's outer growth, they're much more susceptible to cold and wind as well as predators.

If you're thinking of buying and releasing a gallon or so of lady beetles, the California Cooperative Extension warns that this seldom gives suc-

Figure 12.2.
European
earwig. (*Ear-
wigs and
Their Control.*
Division of
Agricultural
Sciences,
University of
California,
Leaflet 21010,
1982.)

cessful control. The insects, which are collected from their hibernating places in the mountains, fly all over the area. "None," says entomologist Walter Ebeling, "will stay in the garden."

Cutworms

These nuisances, the larvae of certain moths, destroy young plants by nibbling at a seedling's base. You can take several tacks to get rid of them. An inverted cabbage leaf set near infested plants will trap the cutworms at night. Handpicking at night, with the aid of a flashlight, is another tactic. If that's inconvenient, ring your plants with an empty pet food or tuna can with both ends sheared off.

Earwigs

Earwigs are not necessarily pests. Actually, most are beneficial. However, sometimes they do eat plant material or ripe fruit. To make sure they're the culprits, check your damaged plants at night with a flashlight.

One of the best controls for earwigs is trapping. Use either rolled-up newspaper, a section of old garden hose, or a length of bamboo. Set the traps around the garden and pick them up every day or two. Knock the earwigs into a pail of kerosene or just crush them.

Gypsy Moth

Gypsy moths are a major forest pest in the eastern United States where, during their peak population years, they defoliate millions of trees. Few of the trees, however, die. Decades of intensive pesticide application have not curbed this insect, which now pervades about 150 million forest acres. Carbaryl is sprayed against gypsy moths, but by killing off beneficial insects the chemical may just be compounding the damage. Pennsylvania and Connecticut have stopped using it against the gypsy moth.

You can protect your garden with banding, as described under "Mechanical Barriers" above. You can also apply the nontoxic *bacillus thuringiensis* to infested trees. Watch for the moth's egg masses and destroy them. They're generally found on outdoor items like garden furniture, toys, camping equipment, and building material.

Incidentally, if you're planning a move to a moth-free state from an infested one, federal regulations require you to have all outside household articles inspected before the move and certified as free from gypsy-moth caterpillars, eggs, and cocoons. Quarantined states include Connecticut, Delaware, Maine, Maryland, Massachusetts, New Hampshire, New Jersey, New York, Pennsylvania, Rhode Island, and Vermont.

You can have your articles certified in one of three ways:

- The shipping company may give you a self-inspection kit that includes an appropriate document for you to complete and give to the van driver.
- You may hire a private pest control company to check your property and certify it.
- Your present state's agriculture department may certify that you're not carrying any gypsy moths with you.

Japanese Beetles

The Japanese beetle is a beautiful insect with bronze wing covers and a metallic green body rimmed by small patches of white hair. It was introduced into the United States in 1916. Without natural predators it quickly fanned out from its arrival point, and is now found in just about every state east of the Mississippi. Although still a garden nuisance, it's not nearly so destructive as it once was, thanks to an extensive federal program of spraying milky spore bacteria, fatal to this beetle. The bacteria persist after application and ultimately spread to other areas.

A daytime feeder, the Japanese beetle is most active on clear, warm days and in bright, sunny areas. These habits make handpicking easy since the insect is sluggish in the morning's cool air.

To trap it, spread a large cloth, before 7 A.M., under infested bushes and shake the insects from the plant. Then gather them up and drown them in warm, soapy water. A commercially available trap called "Bag-a-bug" uses a sex pheromone to lure its victims.

Predators of Japanese beetles include chickens, ducks, turkeys, pigs, grackles, starlings, meadowlarks, cardinals, and catbirds. Shrews and skunks also fancy them.

Snails

This universal botheration was introduced deliberately into the United States by French immigrants as a delicacy. Our palates betrayed us for snails are among the worst ravagers of field and garden. A few species prey on small insects and one, the decollate snail, is a cannibal, feeding on its fellow mollusks, but most are extremely destructive.

Snails are most active at night or in damp, foggy weather. When the thermometer falls below 50°F, they are inert. On sunny, dry days they hide under boards, rock piles, and in low-lying shrubs and dense vegetation. They tend to favor beds of ivy. Long dry spells don't kill them; they just seal themselves up in their shells and can remain dormant for as long as 4 years.

Metaldehyde, the active ingredient in most snail baits, is a strong poison, but the snails are becoming resistant to it. To make matters worse, the bait is often extremely attractive to dogs, who will tear a box open to get at it. Metaldehyde can throw a dog into convulsions, even kill it. Nor is it a thing to keep around if you have small children. The pink granules don't look all that different from some candied dry cereals.

Actually there are so many ways to control snails—yard sanitation, traps, barriers, handpicking—that there's really no need to use poison baits. To get rid of snails nonchemically:

1. Clear your yard of unused boards, stones, and other debris.

2. Repair any leaky outside faucets or hoses.

3. Clear out weeds and other unnecessary foliage so that the soil dries more quickly and makes a less hospitable snail environment.

4. Remove dense ground covers like ivy. (If you must keep the ivy, set your snail-susceptible plants as far as possible from it.)

5. Protect low-growing plants (snails favor them) by ringing them with an inch or two of wood ash, sawdust, lime, sand, or diatomaceous earth. All are very irritating to mollusks. Since these barriers are less effective when wet, replace them after a rain. Spread the irritant around the perimeter of vegetable plots to keep snails out of the whole area.

6. Handpicking is probably the fastest way to clear out snails. Let the soil go very dry and then, late in the afternoon turn on the sprinkler. After dark go out with a flashlight to gather the snails (there'll probably be dozens) awakened by the moisture. Put them in a bag and crush them. Since they are rich in nitrogen and calcium, bury them for fertilizer.

   If the weather is so damp that the soil doesn't dry out, handpick whatever snails you find in the morning. By destroying them, you're destroying their descendants. At first, handpick every day, but as their numbers dwindle a weekly gleaning should be enough.

Salt sprinkled heavily on a snail out of its shell will kill the animal by drawing out all its body fluids. However, salt can damage soil so should only be used on pavement. Stepping on the pests seems simpler if they're on a walkway.

Snail Predators.   Snail predators are many and diverse, from lightning bug larvae to human beings. Among the most voracious are ducks and geese, but garter and grass snakes, box turtles, salamanders, some toads, and an occasional dog (possibly with French poodle ancestry) will also eat them. Rats also feed on them, a good reason to keep the snail population of your garden down.

   If you want to become a snail predator yourself, purge with cornmeal or oatmeal for at least 3 days those that you handpick, find a good recipe and invite your friends for an escargot party!

Snail Traps.   A number of simple traps can control snails in an otherwise clean garden. Figure 12.3 illustrates an easily made board trap, set on runners, that attracts snails by giving them a shady cover. Just be sure

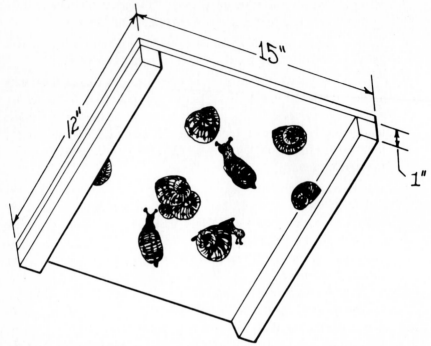

Figure 12.3. An effective snail trap. (Boards on 1-inch risers can be placed in the garden to trap snails. Snails should be removed and disposed of each morning.) (*Snails and Slugs in the Home Garden*. Division of Agricultural Sciences, University of California, Leaflet 2530, 1979.)

to empty it every morning or the snails will crawl off and continue their gorging.

Another effective trap is a saucer of stale beer mixed with a bit of wheat flour to make a stickier brew. Very thin fermented bread dough will also catch them. Raw slices of turnips or potatoes will bring them out and flower pots and grapefruit shells turned rim to the ground are said to provide ideal snail hideaways and, hence, traps.

Sowbugs and Pillbugs

Sometimes called wood lice, these tiny creatures are not really bugs at all. They're related to lobsters, crabs, and other crustacea. For the average person the only perceptible difference between the two is that the pillbug curls itself up when threatened.

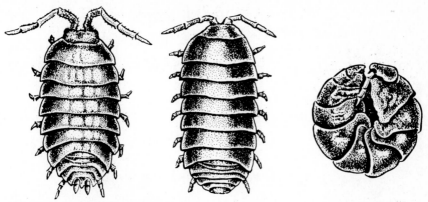

Figure 12.4. Sowbugs and pillbugs–left to right: a sowbug, a pillbug, and a pillbug in the typical defensive position. (*Sowbugs and Pillbugs*. Division of Agricultural Sciences, University of California, Leaflet 21015, 1980.)

Overall, sowbugs and pillbugs are useful scavengers who turn decaying organic matter into soil nutrients. Occasionally, they may injure a young plant.

If too many are in your yard and they're getting into the house, try to eliminate the damp conditions they like:

- Remove all boards and boxes in the garden.
- Eliminate piles of grass, leaves, and other hiding places, especially those next to the foundation.
- Plant shrubs and flowers far enough from the house so that wind and sun can keep the foundation area dry.
- Keep foundation and sills tightly caulked.
- Ventilate crawl space and basement adequately.
- Don't overwater.

Spider Mites

"When you kill off the natural enemies of the pests," says scientist Carl Huffaker, "you inherit their work." Spider mites are a prime example of this. Once minor pests, since the introduction of synthetic organic pesticides, which wipe out mite predators, they have ballooned into a serious problem worldwide. Because its predators—lady beetles, lacewings, flower bugs, and predaceous mites—take longer to reproduce than the spider mite does, killing them off leaves the pest free to reproduce at an enormous rate.

Signs of spider mite infestation are leaf damage, webbing, bronzed or stippled leaves and stems, poor leaf and flower color, fewer fruit, and buds that fail to open.

A dormant oil spray applied before the growing season blocks the mite's population buildup. The United States Department of Agriculture recommends hosing infested plants with a strong spray of soapy water. In warm weather the spray should be applied several times a week. Be sure to reach all parts of the plant, including the undersides of leaves.

Pest Control for Houseplants

Houseplants may be more susceptible to insect pests than shrubs and flowers growing outdoors because, once infested, there are no predators to clean things up. Some preventives for keeping your indoor greenery pest-free are:

- *Don't bring infested plants into your home.* Examine any new ones carefully for signs of infestation.
- *Isolate new plants for a month and watch for pests.*
- *Give plants a tepid bath in soapy water every 2 to 4 weeks.* Use 2 tablespoons of detergent to 1 gallon of water. Rinse in clear water.

Soapy water dips or baths will control aphids, spider mites, and white-flies. Whiteflies, however, soar off a plant as soon as it's disturbed. To prevent this pop a plastic bag over the infested plant, take it outdoors, and wash it well. Repeat this every few days.

Mealybugs are easily killed by a dab of alcohol on a cotton-tipped swab. A weekly bath in soapy water also helps.

Remove scale insects with your fingernails or the tip of a sharp knife. You can also pick them off with a tweezers.

To treat a plant that's heavily infested with any of these pests, cut away the blighted parts and dispose of them with the garbage. Isolate the plant for a few weeks.

Some houseplant problems may not be caused by insect pests at all. Inadequate light, poor air circulation, sudden bursts of cold air, water-sprayed leaves can all damage tender tropicals. So check these factors before treating for what may be nonexistent insects.

Insects Basic to the Natural World

If we can accept insects and other small wildlife as integral to our home's landscape, we'll be more willing to share space with them, more reluctant to call out the troops at the first sign of an insect, beneficial or otherwise. Healthy plants can tolerate considerable damage without keeling over, so before turning to chemical pesticides, try some patience, a bit of ingenuity, regular maintenance, and a willingness to accept a less-than-perfect garden.

PART FIVE

CONTROLLING THE CONTROLLERS

CHAPTER 13

Ants, Spiders, Wasps—Useful but Unwelcome

The ant, the spider, the wasp—each is both a help and a pest. When they burrow in the soil, enriching and aerating it, or when they attack termites and other insects that harass us, ants are our welcome allies. But let us find them swarming over the kitchen countertop after a hard rain and we want them out yesterday. Seeing a spider's web stocked with dead houseflies and gnats, we're grateful that these disease carriers have been dealt with for us. Nevertheless, even a small web in a room's corner sends a careful homemaker flying for a broom. We may be only vaguely aware of what wasps do that benefit us but, "Please!" we say, "let them do it somewhere else."

Scientists who spend their entire lives studying these insects express wonder at their abilities. A spider's web is an engineering miracle, exquisitely suited to providing its builder with food. Some species of wasps make their superbly engineered nests of paper so sturdy you can write or even type on it. Ants set up their social systems millions of years before humans.

However astonishing their talents and however great their benefits, when ants, spiders, or wasps get too close for our peace of mind, most of us want to get rid of them firmly, safely, and as quickly as possible. Let's consider ants first since they're the most likely to move in with us and bring all their coworkers.

Nature of Ants

Ant Societies

Ants are found almost everywhere on earth, from snowy mountain-tops to steaming swamplands. They surpass in numbers all other land animals. Their great underground cities, says naturalist Edwin Way Teale, can outlast a human generation.

Ant societies are organized by castes, each with its well-defined tasks: Queens and males are responsible for the group's continuity, female and neuter workers for its day-to-day survival. Workers forage for feed, keep the eggs and cocoons clean, feed the larvae, and defend the nest. The same keen sense of smell that brings them to the jam you forgot to wipe up yesterday tells them who's a member of the colony and who isn't. In many species an ant that doesn't smell like the rest of the colony is barred from the nest.

If a potential predator shows up, the workers squirt it with formic acid, a form of chemical warfare at which they're experts. They can shoot the acid up to a foot away, a huge distance for a being about $\frac{1}{10}$-of an inch long.

Ants resemble termites in some of their habits. Both species mate in huge flying swarms. Both drop their wings and nest in hidden places—underground or deep within wood or a decaying tree. Both can form enormous communities, up to a quarter of a million individuals in one nest. Egg-laying ant and termite queens are imprisoned for life in their chambers and must be cared for by workers. In contrast to the secretive termites, ant colonies are usually quite visible, with workers hurrying back and forth on important errands.

Whereas termites eat only wood-related product, ants will eat just about anything humans do. They'll devour sweets, fats, starches, grains, meats. When seeking meat, ants, in large numbers, can kill a young bird or chicken.

Ants' Abilities

In earning their living, ants show remarkable adaptability. The harvester ant nips the seeds that it gathers into its nests to keep them from sprouting. After a rain, workers carry the seed to the surface to dry them out and prevent them from molding. Another species, the leafcutter, carries bits of leaves underground to form a fungus garden, which it fertilizes with its own excrement. Ants that tend aphids sometimes build paper shelters over their "cattle" to hide them from predators. They've also developed war to a high art. Winners of a battle carry home the cocoons of the conquered. The hatchlings become slaves of the victors, who ultimately may pay a high price for their victory by forgetting how to take care of themselves.

More Help than Harm

The nuisance that ants can be is relatively slight when weighed against their many benefits to us. Besides enriching the soil, they prey on cockroaches, conenose bugs, the larvae of filth flies, scale insects, mealybugs, and other pests. They will even attack the oriental rat flea, bearer of plague.

In thirteenth-century China, orchardists used warlike ants to guard their orange trees from citrus pests. Today German law protects red ants so they, in turn, can defend the nation's forests. In cotton-growing regions of the United States ants help control the boll weevil. One of their greatest benefits to us is their fondness for mosquito eggs, which can stay in a dry spot for years until a rain or other watering sets them to hatching. Foraging ants often prevent this.

Ants can be good termite insurance, for the two are age-old enemies. Because ants will attack a wingless queen termite searching for a nest site, they probably prevent much termite damage to our homes.

Ant Invasions

Ants are most apt to invade a home after their nest has been flooded. It doesn't matter if the flood is caused by heavy rain or just garden watering. Our son recently repotted some houseplants outside the house and forgot to turn off the garden hose. An hour later, the outside wall was covered with thousands of tiny confused creatures, suddenly homeless. Their scouts soon found a crack or two in the wall and signaled to the whole tribe that delicious tidbits awaited on the dirty dishes in our dishwasher. They also found a dab of dried jam in one of the cupboards. It took us a week to persuade them to go live somewhere else. As we discovered, ousting ants takes ingenuity and patience. Also a sense of humor. We did finally succeed and without chemicals.

Shut Ants Out

Preventing Ant Invasions

It's much better to keep ants out then to try to evict them once they've found their way in. Here are some simple preventives:

- Trim all tree and shrub branches well away from the house since ants can crawl along them and get in through window and door cracks.
- Patch all wall cracks both inside and out.
- Don't overwater your garden, especially near the house.
- Wipe up all spilled foods immediately.
- Rinse dishes well before putting them into the dishwasher; or wash them by hand after each meal.
- Store all foods in tightly closed containers.
- Wipe clean all bottles of jam or syrups, including medicine syrups, right after using them.
- Wrap leftovers, especially cake and cookies, closely in foil or plastic.
- Sweep your kitchen floor daily.

Getting Ants Out

Once ants have made their way into your home, trace them back to their entry point and seal the hole with petrolatum, putty, or plaster, says William Olkowski. Petrolatum gives very quick results but for more permanent effect, use the long-lasting materials.

You probably won't get complete control right away, but persevere. Eventually you'll close off so many of their entries that the few stray crumbs around your kitchen will be too far from their nests to be worth their while.

Vacuum or wipe up any stragglers still wandering around indoors. Be sure to empty the vacuum bag outside or ants may come crawling out of it to pester you all over again.

These energetic foragers can get in through surprising crannies—a crack around a window frame, a circuit breaker with tiny crevices around the wires. We once found them coming through a light switch. A dab of petroleum jelly embalmed the fellow poking its head out alongside the button and kept out any of its mates.

Incidentally, the quickest way to deal with an ant swarm in your dishwasher is to run the machine.

Mint and Other Herbals

Mint and other herbals are often recommended as ant repellents. I've put freshly picked mint and tansy in a kitchen cupboard during an ant invasion. The insects stayed away from that cupboard for about an hour. Not until every sweet spot was cleaned up and every wall crack was closed, did I finally win control.

It's the housecleaning that goes along with spreading the "repeller," says ant expert Robert Wagner of the University of California at Riverside, that really does the job.

Wipe Ants Out

Destroying a Nest Outside

If your efforts in the house are not giving you thorough enough control, you can destroy the nest outside. Pour boiling water or hot paraffin down the nest entrance. Since ants can survive for days under water, flooding the nest with cold water won't do the trick.

Insecticides for Ants

Baits containing arsenic, once widely sold for ant control, can still be found on garden shop shelves as well as in homes. Don't use them. Arsenic combines readily with common elements like hydrogen and oxygen to form extremely poisonous substances. It also remains indefinitely in the ground, never decomposing.

Other chemicals can also cause harm. Recently a young woman developed a grave form of anemia. Her doctors suspected the cause to be a household pesticide she'd sprayed heavily in her small kitchen to kill a stubborn ant infestation. If your problem is so bad that you're considering using poisons, you'd be much better off to hire a competent pest control operator to clear up the situation quickly.

Future Controls?

University of California scientists have found that soldiers of several species of desert termites produce a powerful ant repellent. The biologists hope to reproduce the repellent synthetically and foresee millions of U.S. homes ringed by the substance while the ants march peaceably away. Since ants kill termites, keeping them away from a house could be a double-edged sword. Which would you rather take a chance on—ants or termites?

Spiders

On a sunny day in 1969 a group of citizens gathered near the entrance to the city park in the small town of Sierra Madre, about 20 miles northeast of Los Angeles. Television crews bustled about, focusing their cameras on a brown metal object atop a pedestal.

The center of the fuss was a statue of a spider. It may be the world's only monument to a spider, but this species, the "violin" spider—one of the most dangerous—had put that small community in the spotlight, so it well deserved the honor.

The statue marked the first time an infestation of the violin spider, also known as the South American brown spider, had been seen on the west coast of the United States.

Spiders in Legend

Spiders have fascinated people since time immemorial. The Greeks told of Arachne, so proud of her beautiful weaving that she challenged the goddess Athena to match her fabrics. Offended, the goddess destroyed the mortal's work. In despair, Arachne hanged herself, whereupon the compassionate Athena changed her into a spider.

The Navajo say that the spiders taught them to weave and, to this day, their grateful weavers leave a tiny hole in their blankets to recall the entrance to a spider burrow.

And what English-speaking child hasn't heard of Robert the Bruce who, discouraged and facing defeat in his war against the English, gained new resolve while watching a spider's persistent efforts to build a web in a difficult corner? When it finally succeeded, he decided to try one more time. He too succeeded.

Potent Insect Control

Although a few species, like the brown violin, are a danger to humans, spiders are much more helpful than most of us give them credit for. They are among nature's most effective insecticides. According to Dr. Willis Gertsch, former curator of the American Museum of Natural History, spiders are probably far more important than birds in the number of insects they destroy. Lots of people are sentimental about birds, but who melts with tenderness over an eight-legged spinner? Yet, these rather stupid, nearsighted, extremely timid creatures are essential to farms and forests. They help control cottonworms, gypsy moths, pea aphids, among many other destructive pests. They're especially important in cultivated cotton and cornfields, where these crops are most vulnerable to ravenous insects. Quickly overrunning the fields when the pests emerge, spiders wipe out larvae and adults.

They're also most useful in cities. In 1925 crab spiders stopped a huge infestation of bedbugs in Athens, Greece. The same species is often found in New York City's food warehouses, preying on stored-food pests. Other spiders kill moths and grasshoppers. The wheel net spider, which destroys an average of 2000 insects in its 18-month life, is used in Germany to protect forests.

Nature of Spiders

Most Spiders Are Harmless

Spiders carry no known diseases. Indeed, Italian peasants believe that cobwebs in the stable are directly related to cattle health. There's probably more than a little truth to this because trapped in the skeins are stable flies, houseflies, gnats, mosquitoes, and other health menaces.

Most spiders pose no threat to humans. The fangs of the garden spider are so tiny they can't pierce our skin at all. If spider venom is needed for an experiment, the creature often has to be forced to bite, sometimes with electric shocks.

And they have some remarkable talents. Crab spiders can change their color to blend with the flower petals concealing them from their prey. The bolas spider is the original cowboy, throwing out a line weighted at the end with sticky liquid silk to rope in its prey. Like insects, spiders have the ability to regenerate an amputated leg.

Probably the most marvelous thing about them is the most common—

their webs. Whether as delicately elegant as the garden spider's or as messy as the black widow's, it is an engineering marvel, precisely suited to catching its builder's daily bread. Meticulously built, the garden spider's orb web takes about an hour to weave, a task the methodical creature does every morning since the old one is spoiled by dead insects, dust, and fallen leaves. The black widow's untidy home always has a tunnel where the timid creature can hide.

Spiders' silk is as remarkable as their webs. Once used in ancient China as sewing thread, the fibers range from 20 millionths of an inch in thickness to 1 millionth of an inch. Their tensile strength is said to equal that of fused quartz, yet each fiber can be so fine that human technology can't match it. The fibers make excellent markers for surveying and laboratory instruments. The cross hairs of the Norden bombsight used by the Allies in World War II were made from the silk of the black widow.

Beware of Widows and Violins

Although most spiders are harmless and generally helpful for human purposes, two fairly widespread species are dangerous: the black widow and the South American brown, or violin, spiders. Both are shy and nocturnal. See Figures 13.1 and 13.2. Those most likely to suffer a severe reaction to their bites are very young children, the elderly, and people in poor physical condition.

The chart on page 221 summarizes these spiders' characteristics and their hiding places. Be careful around *any* place or thing that has not been disturbed for a long time. Picking up a plumber's helper once, I was startled by a sooty widow scurrying out from the plunger. (She was probably more scared than I.)

The Black Widow Spider

Despite its fearsome reputation, the black widow spider kills many fewer people than do snakes. Between 1716 and 1943 there were 1300 cases of black widow bites in the United States, with fifty-five deaths recorded. Each year, by contrast, about 1500 poisonous snakebites are reported in the country and, on the average, seventy-five of the victims die. In other words, more people die from snakebites in this country in one year than died here from black widow bites in 227 years.

We can clean up their reputations a bit but, with venom fifteen times more powerful than that of a rattler (but fortunately in comparably minute

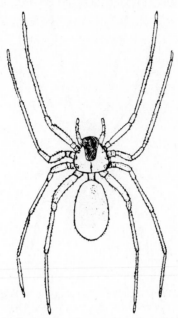

Figure 13.1. The recluse, or violin, spider. (*Household and Stored-Food Insects of Public Health Importance and Their Control,* Center for Disease Control, Home Study Course 3013-G, Manual 9, 1982.)

quantities), a black widow is nothing to take lightly. Control and precautions are essential. English zoos keep this species on exhibit, but when the international situation darkens, the colonies are destroyed so that no stray bomb can release them.

The Bite of the Black Widow

"The bite of the black widow spider need never be fatal," says entomologist Walter Ebeling, "if treated promptly by a physician." However, *anyone who's been bitten should receive antivenin at a medical facility as soon as possible.* Since the venom acts very quickly, cutting the wound open and sucking it does little good.

Try to kill the biting spider without squashing it beyond recognition, and give it to the doctor. Treating for the wrong spider bite could be harmful.

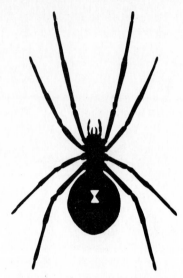

Figure 13.2. The black widow spider. (Courtesy of James M. Stewart of the Centers for Disease Control, Center for Infectious Diseases, Atlanta, Georgia.)

Victims under 16 or over 60 years old or with high blood pressure should be hospitalized right away. Young children should be watched closely during their first 10 hours of hospitalization.

Symptoms of a black widow's bite are:

- Intense pain at the wound site.
- The pain then moves to the abdomen and legs.
- The abdomen becomes rigid and boardlike.

The bite can also cause:

nausea and vomiting	shock
faintness and dizziness	paralysis
tremors	speech disturbances
loss of muscle tone	fever

Characteristic	Black Widow Spider	Violin Spider
Color	Shiny black; sometimes gray	Varies from light fawn to dark brown
Size	Body, ½-inch; globular abdomen	Body, ¼- to ½-inch; long legs
Markings	Usually red hourglass on abdomen	"Violin" on back, scroll toward rear
Signs of infestation	Webs; spider itself	Long-legged, shriveled cast skins; webs; spider itself
Webs	Tangled, coarse, on or near ground in dark places; can trap a small mouse	Above floor: white cottony with bluish support strands; usually with egg cases near floor: long, wide, multilayered bands
Hiding places	Barns, henhouses, garages, cellars, outdoor privies, gas & water meters, angles of doors & windows, behind shutters, woodpiles, rubbish, old tires, grape & tomato vines	Boxes, among papers, under rocks, in old clothes, undersides of tables & chairs, corners, behind pictures, at tops of window shades, under tree bark, in bird skeletons

In addition to the antidote, treatment may consist of mild sedation, muscle relaxants, and hot baths. *In no case should the patient be given any alcoholic drinks.*

Individual reactions to spider bites vary greatly. Untreated bites can lead to such serious complications as erysipelas (a skin infection), tetanus, and cerebral hemorrhage.

It may be some comfort to know that "despite its severe symptoms, spider bite poisoning is, in a majority of cases, a self-limiting condition, and generally clears up spontaneously within a few days,'' according to Dr. Emil Bogen.

The Violin Spider

Two kinds of violin spiders, the recluse and the *laeta,* are found in the United States. The recluse is fairly widespread in the midwest; its bite, though serious, is rarely fatal. The more dangerous *laeta* first showed up in the United States in 1965, when a large infestation was discovered at Harvard University in Cambridge, Massachusetts. Five years later the species surfaced in southern California. They've also been seen in Ontario and British Columbia. The species is native to South America, where up to 1968, it had bitten 400 times and caused at least thirty-five recorded deaths.

The California infestations were spread in cartons of merchandise. One clutch of them was in clothing that was slated for a church rummage sale.

The Bite of the Violin Spider

Although the violin spider's venom can have very serious consequences, the animal is reluctant to use its fangs. It will attack, however, if caught in clothing one has just put on or if rolled on in a bed.

Today doctors use corticosteroids to treat violin spider bites, and with good results, but *it is essential that anyone bitten by this spider get to a physician as soon as possible*. Meanwhile put ice on the wound. If possible, kill the spider without mutilating it and take it along so it can be positively identified.

The poison causes a severe reaction at the wound site. Effects of violin spider bites:

- Gangrene develops around the puncture.
- The skin eventually sloughs off, leaving muscle and tendons exposed.
- Healing can take up to 8 weeks.
- Heavy scarring can require plastic surgery.
- Where the spider has injected the maximum amount of venom, the victim may suffer a systemic reaction, including blood and kidney problems.

Shut Spiders Out

Because spiders can shut down their respiratory systems much better than insects, they're relatively immune to most home pesticides. Prevention and mechanical controls are much more effective, and both are largely matters of good housekeeping and common sense.

Prevention Outdoors

1. Keep your garden cleaned up.

2. Break down any untidy webs you find.

3. Plant or trim your shrubs well away from the house to allow sun and wind next to the building.

4. Make sure the basement and any space under porches are dry. Keep the foundation well caulked.

5. If your children have a tire swing, paint the inside of it white so any spiders in it will be clearly visible. Look it over carefully before it's used and, as your youngsters get older, alert them to this danger.

6. Before using an outdoor privy, take a stick in with you and run it under the seat to knock off any creatures underneath. (This is a favorite hideout for black widows.) Teach your children to do this. A bite on the genitalia is very serious and men are the most frequent victims. If you're spending an extended period in a rustic area without indoor toilets, paint the underside of the privy seat with creosote or crude oil to repel spiders.

7. Handle garbage as recommended in Chapter 6.

8. Dispose of pet droppings promptly so that spiders will look elsewhere for their prey.

9. Use yellow bulbs for outdoor lighting. They attract fewer insects and, so, fewer spiders.

10. Keep storage areas reasonably tidy.

11. Always wear sturdy work gloves when cleaning garages, tool sheds, and garden houses, or even when just rummaging about in them.

12. Carefully inspect an old jacket or sweater or pair of coveralls hanging in the garage or tool shed before putting them on. These

may be a great convenience for whoever does the yard work, but spiders often slip into clothing hanging on a wall.

13. If you suspect that black widows or violin spiders are on your property, put on sturdy closed shoes and look for them at night with a flashlight. Step on any you find.

Prevention Indoors

Some of the same measures that keep out other unwanted wildlife will suppress spiders in your house. The British have a saying, "There are three spiders in each room of every house no matter how houseproud the owner."

1. Tight screening of windows, doors, and vents, keeping out spiders' prey, will make sure the number in your house doesn't go above that number.

2. Inspect all firewood, plants, and cardboard cartons before bringing them inside just as you would for cockroaches and carpet beetles.

3. Sweep behind washers and dryers regularly to oust potential troublemakers since these appliances give spiders ideal nesting conditions, dark and moist enough to attract unwary prey.

4. Move heavy furniture at reasonable intervals so you can clean beneath it.

5. Periodically remove and clean curtains, pictures, and luggage.

6. Consider rearranging your furniture. (In Chile, where violin spiders are a serious problem, scientists recommend keeping a bed at least 8 inches out into the room so that wall-crawling spiders can't get to it.)

7. Vacuum up any webs and egg cases and empty the bag outside as soon as possible in a closed garbage container, preferably set in the hot sun.

8. Shake out any garments and shoes unworn for a long time before putting them on and look down into the shoes' toes before slipping into them.

9. Brush off, never swat, anything you feel crawling over your arm, the back of your neck, or face when you're in bed at night. Swatting guarantees a bite, possibly venomous, while brushing

catches the small explorer by surprise and removes it before it has a chance to do any harm.

10. A prime rule in preventing spider bite is: Never thrust your hand or foot where your eye can't see. A corollary is: Never poke your bare hand up a chimney flue.

Wasps

On a recent hot summer's day a retired couple decided to spend a few hours at a nearby mountain lake. While unpacking their lunch, they became the center of attention of an aggressive clan of wasps who attacked their food and invaded their picnic basket. The two quickly packed up and left the area. Bathers at the same lake were also fending off the irascible buzzers.

You may not class wasps with rattlesnakes, but before a vaccine was developed, wasps killed more people than rattlers did. And a wasp did it a lot faster. Eighty percent of those fatally stung died within a half hour of the insect's attack, while only about 17 percent of rattlesnake victims succumbed in under 6 hours.

So aggressive are some species of wasps, they've even been recruited for war. The Vietcong used these "winged guerrillas" to stop the advance of South Vietnamese troops.

Despite their unpleasant dispositions, wasps are helpful to us. They kill filth flies, black widow spiders, and many agricultural pests. Without one particular miniwasp, the Calimyrna and Smyrna fig trees would not bear fruit, and one of California's major industries would fold up.

The galls, or swellings, that wasps make in some trees provide us with excellent inks and dyes. Aleppo oak ink, produced by a wasp gall, was once required by the United States Treasury, the Bank of England, and other government agencies to be used on official documents.

Nature of Wasps

Some Dangerous, Some Not

The jury may still be out on wasps as friend or foe, but it's best to keep them at a healthy distance. To decide whether the wasp you're nervously eyeing is a real menace, you need to know something about the family as a whole. The more than 20,000 species can be roughly divided into two groups—solitary and social.

Solitary Wasps Usually Harmless

Solitary wasps can nest underground, in a nail hole, or in a pithy plant stem. There they lay their eggs and stock the burrow with paralyzed insects on which the larvae feed.

Solitary wasps are usually no threat to us. Although some are armed with frighteningly long stingers, they use them almost exclusively for paralyzing the insects they drag into their nests. They will often specialize in one kind of prey; for example, the cicada killer, so large it's also called the king hornet, preys only on cicadas. Another species preys exclusively on grasshoppers, another only on flies.

Watch Out for Social Wasps

Social wasps, intent on protecting their nest, are much more apt to string humans they perceive as intruding. Depending on the species, wasps' nests, which can measure up to 6 inches across, are built in leaf litter, clumps of sphagnum moss, in trees, under eaves, on the ceilings of garages, or underground. Aerial colonies tend to be smaller than those underground.

In cooler climates, the colonies die when the temperature falls below freezing. The only survivors are fertilized queens, who hibernate until spring, sometimes in human homes. If you find a wasp at a window in spring, either let her out or swat her with a flyswatter or you'll soon have a nest inside.

Bees, ants, termites—all nesting insects—swarm when they're ready to start new colonies. Wasps do not. New communities are founded in the spring by the emerging queen. The new colonies never reuse an old nest. Once empty, it's no danger to you. However, the insects may build a new one nearby. Incidentally, according to the California Cooperative Extension, tree-nesting wasps are beneficial. Unless the nest is located where someone is apt to come upon it unexpectedly and startle the wasps, it should be left alone.

Wasp Stings

Most wasps, says wasp authority Howard Ensign Evans, have to be handled pretty roughly before they'll sting, and, he adds, "to be stung by one is about as likely as being struck by a falling acorn." Only female wasps have the sting. Unlike the bee, which dies after stinging once, the wasp can needle a victim repeatedly.

Wasps are more irritable on rainy days; sunny, warm weather keeps them in a mellower mood. They're antagonized by bright colors and are more apt to go after someone wearing vivid clothing. In the fall, with their colonies dispersed and no nests to defend, the insects are sluggish and not so quick to fly off the handle.

The effects of a wasp's sting can vary from a mild prick to severe local pain. To treat a sting:

1. Wash the area with soap and water (the sting may be dirty.)
2. Apply an ice pack or cold compresses of witch hazel to relieve swelling and pain.
3. An antihistamine cream such as Benadryl, applied within 20 minutes of the stinging, is said to be effective.
4. Baking soda also helps.

The Extremely Sensitive

Wasp and bee stings are life threatening for some people. Severe swelling in other parts of the body, nausea, vomiting, and cramps; difficulty in breathing; dizziness; hives; unconsciousness are all possible complications. Since 1959 vaccines have been available that can immunize most of the intensely sensitive. In addition to the vaccines, an antidote kit is available with a doctor's prescription. It consists of adrenalin and an injector to block the reaction quickly. If you or your child has had a strong reaction to an insect bite, consult a competent allergist about immunization.

Scientists at Berkeley's Center for the Integration of Applied Science recommend that those allergic to bee and wasp stings, in addition to having themselves immunized, follow these guidelines:

- Avoid wearing perfumes or perfumed substances like hair spray, suntan oil, and aftershave lotion.
- Wear white or light-colored clothing
- Walk around, not through, masses of flowers or flowering shrubs.
- Avoid outdoor areas of strong food odors such as barbecues and open garbage cans.

Shut Wasps Out

Sanitation Gives Good Wasp Control

To hold down the number of wasps in their vicinity, all households should

1. Keep outside garbage cans tightly covered.
2. Dispose of any overripe or rotting fruit in plastic bags set inside garbage cans.
3. Fill up any abandoned rodent burrows on your property as wasps often nest in these.
4. On picnics have a small can of tuna or pet food with you. Set it opened some distance from your party to attract scavenging wasps away from your meal. If one comes near, stand still. Don't swat at it. If possible, put a handkerchief between you and the insect.

Wipe Wasps Out

Destroying a Wasp Nest

Wasp nests, whether above or below ground, that are near areas of activity like entranceways and children's play equipment should be destroyed.

Underground Nest

Destroying a ground nest demands great care and considerable preparation. The safest way is to have a certified pest control operator do the job. If you must do it yourself, recruit a helper and follow these steps:

- Have a half gallon of kerosene or diesel fuel (*never use gasoline*) ready in a wide-mouthed container.
- Also have a shovel and a good-sized cloth.
- Both of you should wear protective clothing, including rubberized gloves and hats with bee veils. Slick-surfaced garments are safer than cotton or wool, which can be easily penetrated by the wasps' stingers.
- Tie your trouser legs and jacket cuffs over your boots and gloves.

- Plan to do the job after dark or very early in the morning, when the insects will be in the nest.
- Approach the nest very quietly so as not to alert the highly sensitive occupants.
- Pour the fuel into the nest opening and cover the hole immediately with the cloth to hold in the fumes. Have your helper shovel dirt over it.
- *Do not ignite the fuel.* The fumes alone will kill the insects very quickly.
- Leave the area immediately.

This method is preferable to destroying the nest with insecticide since the chemical works much more slowly than kerosene or diesel fuel, and you're much more likely to be stung.

Aerial Nest

The safest way to destroy an aerial paper nest is with a pressurized bomb specifically designed for wasps.

1. Do not use a general household insecticide. These work too slowly and you run a high risk of being attacked.
2. Be sure to follow label directions closely.
3. Stand 8 to 10 feet away and let the bomb shoot its stream of liquid into the nest. If you can get the stream right into the nest opening, so much the better, but just soaking the paper cover will also kill the colony.

Mud Daubers

One aerial nest that's easy to control is that of the mud dauber. These nests have only one occupant, an egg-laying female. Wait until she's either just leaving or just entering the nest and kill her with a fly swatter. Then destroy the nest. During the winter you can safely remove the nest without insecticides because only nonflying immature forms of the insect will be present.

Reducing Wasp Populations

Scientists of the United States Department of Agriculture recommend a "fish-wetting agent-water" trap as an effective way to reduce

Figure 13.3. Wasp trap. (Raw fish with sides cut to expose flesh, suspended above a pan of water to which a wetting agent has been added. Yellow-jacket workers attempt to fly away with a large piece of flesh, fall into the water and drown.) (*The Yellow-jackets of America North of Mexico.* Washington, D.C.: United States Department of Agriculture Handbook 552.)

the number of wasps in the neighborhood. A raw fish is suspended, as in Figure 13.3, above a tub filled with water plus some detergent. The skin on the sides of the fish is cut open to give the insects ready access to the flesh. Yellow-jacket workers usually take pieces that are too heavy for them, fall into the water, and drown. If dogs and cats steal the bait, cover the trap with chicken wire or hardware cloth.

CHAPTER 14

Pesticides: Only as a Last Resort

It's a pleasant Saturday in June and the family is tackling its pest problems. Dad sprays the basement for roaches and ants while Mom treats the woolens for clothes moths. A pest control operator went over the foundation yesterday for termites. Finished with the basement, Dad douses his roses with a chemical to kill thrips and aphids, getting some of the stuff on his arms and clothing. He has no idea of how much he's used. Then he applies another chemical to the lawn to get rid of lawn moths.

Junior, sick of seeing the dog scratch, dusts her all over with flea powder. He wears no mask. Through all this the family's 2-year-old crawls around, "mopping up" at each operation. As we leave this busy group, the baby's trying to work the spray button of a pesticide dispenser Dad's left lying around.*

Pesticide Use

A slight exaggeration? Perhaps, but not too far from the truth. In a study published in 1979, the Environmental Protection Agency (EPA)

---

* Adapted from Frank Stead, "Pesticides in Relation to Environmental Health." *California Vector Views*, Vol. 11, No. 1, p. 3, Jan. 1964.

found that more than 90 percent of all U.S. households used pesticides in the yard, garden, and house. Nearly 84 percent used these chemicals indoors two or four times more than they used them outdoors.

Careless Use of Pesticides

Most of us are woefully ignorant when selecting and using a pesticide properly. Asked what they're using, people usually answer, "Raid" or "Black Flag," equating the chemical with its trade name.

Their only source of information may be the retail salesperson in a nursery or hardware store, and many clerks are as uninformed as their customers of the real nature of what they're selling. Most customers rely on ads or pick the most eye-catching container on the shelf.

The EPA has found that those who bother to read labels (fewer than half the buyers) are usually just looking for instructions on how to apply the chemical. Very few study the ingredients or antidotes. Even fewer check with a county agent or college biology department to make sure they've correctly identified the pest they're trying to control.

And when consumers get these potentially harmful products home, they use them as casually as they bought them. They'll turn a roach spray on flies, ant sprays on fleas; they'll release a surface spray into the air, set off a patio fogger on a windy day, hang a pesticide strip in an enclosed van.

Not surprisingly, three-quarters of all pesticide poisonings in 1981 happened at home, and children under age 10 were the most frequent victims. When doctors see a family ill from organophosphates (a group of common antipest compounds), the babies are always the sickest, says Dr. Susan Tully, director of a major southern California emergency room. These substances tend to settle on floors and carpets, where crawling infants easily ingest them. Because they're small, children react much more severely than adults to toxic substances.

There are only four ways to prevent pesticide accidents:

- Don't use chemical pesticides at all.
- Learn as much as you can about any that you do use.
- Apply those you use with great care.
- Use the smallest amount that will do the job.

Usefulness of Chemical Pesticides

Pesticides are not the root of evil that some people believe. They keep us well fed. Basic foods like wheat, apples, tomatoes, carrots, and sweet corn could not be grown practically or economically without chemical pest controls. They also help to keep us healthy. Yellow fever, once a major killer in the southern United States, is almost unknown today, thanks in good part to public health authorities' spraying of mosquitoes. Millions of children around the world are growing up free of malaria and other insect-borne diseases because of the wide use of DDT in developing countries. Despite serious mistakes in the early disposal of DDT, it is estimated that, in the first 28 years of its existence, DDT prevented at least 20 million deaths and 200 million cases of disease.

History of Pesticide Development

Modern pesticide use began in 1867, when Paris green, a combination of arsenic and copper, stopped the Colorado potato beetle's devastation of the U.S. potato crop. Around the same time, Bordeaux mixture, a blend of copper sulphate, lime, and water, was developed in France as a spray for grape pests. By the early twentieth century, fluorine compounds and botanical insecticides had come on the scene. The door to effective suppression of unwanted insects had been opened.

But the new substances also brought some worries. By the 1920s the United States government was limiting the amount of lead and arsenic residues permitted on foodstuffs. And as early as 1912, long before the development of modern synthetic organic pesticides, scientists were noting insect resistance to chemical poisons. Among household pests that quickly show immunity, cockroaches, flies, and mosquitoes top the list.

To complicate matters, chemical pesticides usually kill helpful insects like bees and, in some cases, cause overwhelming surges in what were once minor nuisances like spider mites. Although hundreds of pest species are now immune to some of our most powerful insect poisons, only four of the pests' natural enemies have developed any resistance to chemicals. The chemicals are helping our foes and hurting our allies.

Characteristics of Chemical Pesticides

There are 1400 kinds of pesticides in the United States, combined in 30,000 different ways. Some are termed "restricted use" substances, meant to be applied only by licensed pest-control operators against specific pests. Others are "general use" products, considered by the government to be safe enough for use by the untrained public. All pesticides, no matter what their category, are toxic.

Chemical Categories

Chemically there are five categories of pest poisons: chlorinated hydrocarbons (sometimes called organochlorines), organophosphates, carbamates, botanicals, and inorganic compounds. Each group includes deadly toxins and some that are fairly safe for humans *if used with care*. For example, heptachlor and chlordane, both dangerous, are chlorinated hydrocarbons. So is methoxychlor, generally considered one of the least risky pest controllers. Parathion, an organophosphate, can kill or maim for life; one of its relatives, malathion, is safe enough to be sold to the general public. Strychnine is a botanical and so are the pyrethrins, the safest of all pesticides. Deadly arsenic is an inorganic compound, and so is the relatively mild boric acid.

The organochlorines, the organophosphates, and the carbamates all interfere with nerve functions. Some build up in mammalian tissues and eventually disrupt kidney and liver functions. Pyrethrins cause allergy and boric acid can bring on digestive disorders. None of these is something to use nonchalantly.

Compounding the risk, the vehicle in which the substance is dispersed, usually a petroleum distillate, can be more toxic than the pesticide proper.

Pesticide Interaction with Other Substances

Scientists understand very little about the interaction of pesticides with each other or with drugs and food additives. Tranquilizers, for example, are a form of chlorinated hydrocarbon and, like the larger group, act on our nerves. The relatively safe pyrethrins, which may only stun an insect, are often combined with other poisons to make a much more toxic product.

Pesticide Development: Slow, Expensive

No pesticide can be sold in the United States until the Department of Agriculture and the Environmental Protection Agency have granted it registration, and their scrutiny takes 6 to 8 years to complete. One major firm reports that of the 4200 new compounds it tests each year, only 2 get even halfway through the screening process.

These products are expensive to develop. A company may have invested anywhere from $8 to $10 million before the item ever appears on market shelves. With up-front costs so high, pesticide manufacturers naturally push hard to sell their output. And householders, looking for easy pest control, are eager buyers—to the tune of $2.3 billion a year.

How to Use and Store Pesticides

Identify the Pest Correctly

Despite their complex chemistry and hazardous nature, it is possible to choose pesticides wisely and use them safely. Your first step is to identify your target correctly. Do you really have silverfish or are those small crawlers harmless psocids? Do you have an invasion of earwigs or a kind of cockroach? Unless you're positive that you know the species, catch one in a small jar or box and check it against a book on insects at your public library. If you don't succeed there, take it to the biology department of a nearby college or high school. If these are a dry creek, consult your county agricultural agent, a certified pest control operator, or a retail nursery.

Choose the Right Pesticide

Reading pesticide labels is the most important time you can spend in chemical pest control. Consider only those products that mention your target pest. It is illegal as well as unwise to choose a product designed for one type of pest for use against a totally different species.

Label Warnings

All poisonous substances in the United States must carry signal words alerting consumers to their danger. The degrees of potential harm indicated are as follows:

Danger—Poison imprinted over a skull and crossbones. These are deadly. They kill quickly and violently. Ingesting anywhere from a tiny pinch to a teaspoonful is fatal to the average adult. They should rarely if ever be used.

Warning. Moderately toxic. Ingesting anywhere from a teaspoonful to a couple of tablespoonfuls will kill a normal person. The label may also say, "Harmful or fatal if swallowed." These should be used with extreme care.

Caution. Somewhat less dangerous but still toxic. From 1 ounce to 1 pint can be fatal. Use very carefully.

Keep Out of the Reach of Children. The least poisonous. It could take as much as a quart to finish off an average-sized adult.

Use Toxic Substances Safely

- Choose the least toxic substance that will do your job.
- If you plan to use the poison indoors, make sure the label states that it's safe to do so.
- Never use an indoors one that is designed for outdoors.
- When you buy a pesticide for food plants, make sure the label says that it can be safely applied to the plant you're cultivating.
- If the label says to wear goggles, protective clothing, and a respirator while applying the product, do it, even if you have to buy the gear especially for the job. The equipment will cost less than a hospital stay, time lost from work, or damage to your health.
- Don't endanger other humans, pets, wildlife, livestock, beneficial insects, and ground water. Such hazards are often listed on the package.
- To minimize the chance of packages deteriorating and developing leaks, buy only enough to do the job currently at hand, or only the amount you're sure you'll use in a few months.

The following chart of relative pesticide toxicities is based on one developed by the University of California Cooperative Extension. You

Most Dangerous

aldicarb	fensulfothion	parathion
carbofuran	fonophos	phorate
demeton	methamidophos	schradan
disulfoton	mevinphos	TEPP

(Continued)

Dangerous

aldrin	dinitrocresol	nicotine
bufencarb	dioxathion	paraquat
carbophenothion	DNOC	PCP
chlorpyrifos	endrin	penta (PCP)
DDVP	EPN	phosalone
dichlorvos	methylparathion	phosphamidon
dicrotophos	mexacarbate	propoxur*
dieldrin	monocrotophos	

Less Dangerous

akton	diazinon	lead arsenate
azinphosmethyl	dicapthon	lindane
binapacryl	dichloroethyl ether	metam sodium (metham)
BHC	dimethoate	naled
chlordane	dinobuton	oxydemeton methyl
chlordimeform	endosulfan	phosmet
coumaphos	ethion	toxaphene
crotoxyphos	fenthion	trichlorfon
crufomate	heptachlor	

Least Dangerous

aminozide	diquat	ronnel
captan	malathion	rotenone
carbaryl	maneb	TDE
chinomethionat	naphthaleneacetic acid	temephos (temophos)
2,4-D	oxythioquinox	tetrachlorvinphos
daminozide	Perthane	tetradifon
DDT	propargite	
dicofol	quinomethionate	

* Widely used against roaches.

should check the ingredients of any antipest product you're thinking of buying against this list. The information is based largely on animal tests. Not listed are the pyrethrins, warfarin (used in rodent control), and phosphorus. The pyrethrins have caused no deaths, though some people are allergic to them. Warfarin, according to Dr. Sue Tully, in the amounts that kill a rodent cannot permanently harm a human being. Phosphorus is another story. This deadly poison, in the form of a yellow paste, is brought into the United States from Latin America, where it's used in rodent control. It may be called *Electric Paste, J-O Paste, Paste Electrica, Rough-On Rats, Stearn's Electric Brand Paste*. To a child it looks like peanut butter. The overall mortality rate from yellow phosphorus pastes is 42 percent, says Dr. Margaret McCarron, who has studied this substance. *There is no antidote for phosphorus poisoning.*

Select Child-Resistant Packaging

Given a choice between an easily opened package and one that requires more effort, always choose the latter if you have children. Unlike drug manufacturers, pesticide companies are not legally required to offer their customers child-resistant packages. Ideally, such closures should be compulsory on all hazardous substances, but the fact remains that they aren't.

Asked if he thought pesticide packages were safe, Bernard Portnoy, associate director of the Los Angeles County–University of Southern California Medical Center's pediatrics pavilion, answered with some vehemence, "They're not safe at all. They're always colorful. Think about how Raid is packaged and how it's advertised. It's something to *play* with. It sprays easily. It comes in beautifully colored cans."

Corinne Ray, director of the Los Angeles Poison Information Center, adds, "Some of them you just poke a hole in the top and remove a piece of tape. It's impossible to reclose such a container securely. Once opened," she continues, "a pesticide package should have a very child-resistant lid on it."

What exactly is a child-resistant package? It's one that 200 children are given 5 minutes to open by the Consumer Product Safety Commission. Eighty-five percent of the youngsters must fail to open the package before being shown how; 80 percent must fail after being shown. Parents

Figure 14.1. "Mr. Yuk," either threatening or nauseated, warns nonreaders of a container's dangerous contents. Poison center telephone number saves precious time in case of accidental ingestion. (Insignias from Los Angeles County Medical Association Poison Center; Children's Hospital of Pittsburgh.)

should lobby their representatives demanding these standards for pesticide packages. Write to the manufacturers and let them know that you won't buy their product unless it's offered in a child-safe container. They will pay more attention to you than to a government agency.

Only buy the products that a child can't open. Says Dr. Sue Tully, "If Black Flag makes several kinds of containers and the child-proof one sells the best, you'd better believe they're going to keep making that one."

Contact your nearest poison information center or hospital emergency room for stickers like those illustrated in figure 14.1. Called "Mr. Yuk" by some, they give vivid warning of the harmful nature of a container's contents.

Transport Pesticides Safely

Be aware of safety before putting a pesticide into your car. Never carry toxic substances in the passenger compartment. Make sure the containers can't fall or be knocked over by a sudden stop. Protect paper bags from sharp objects, and both bags and boxes from moisture.

Safe Pesticide Use Outdoors

When applying pesticides indoors or out, safety must always be your first concern. The following precautions are suggested by the United

States Department of Agriculture and/or the University of California Cooperative Extension:

- Work in an area with plenty of light and ventilation.
- Read the label's instructions thoroughly *before* opening the container.
- Use a sharp knife or other cutting tool to open paper sacks. *Never* tear them open.
- When opening a container of liquid, hold it well below eye level, with your face turned away from and to one side of the cap or lid.
- If using a concentrate, wear rubber gloves while mixing the brew because you're handling it in its most dangerous form.
- Mix only the amount needed for the immediate job.
- Do not store mixed pesticides.
- Before filling the applicator, run it with clear water to check it for leaks, weakened hoses, loose connections, pinholes in the tank. Repair any of these or replace the equipment.
- Use the product only for pests and plants specified on the label.
- If it's a breezy day, postpone the application until the wind dies down. Always stand upwind of the applicator so that any unexpected puff of air blows the chemical droplets away from you.
- Do not apply pesticides near wells or cisterns, or on lawns when people or pets are around.
- Water a lawn after treating it for soil pests to wash off the pesticide from the grass so there will be less hazard to children and pets.
- Keep children and animals away until the spray has dried.
- Avoid exposing bees. They are extremely sensitive to some pesticides, going berserk and stinging anything in sight.
- Don't let pesticides drift to your neighbors' property.
- Don't harvest pesticide-treated food before the interval specified on the label.
- *Don't smoke while spraying.*

Equipment Cleanup

- Follow label directions.
- Rinse your equipment at least three times with clear water.

- Drain the rinse water on unpaved ground in a sunny area where it will not contaminate food plants or animals.
- Never pour rinse water down a plumbing drain or toilet.

Safe Pesticide Use Indoors

- Assume that a prepared spray is flammable unless the label clearly says it is not.
- Turn off all gas pilots before starting to spray.
- Before spraying or dusting, clear the area of children and pets.
- Remove or cover all food. Take pet food and water dishes away.
- If using a surface spray, make sure it falls only on the surface you need to treat. Don't aim it into the room's air.
- Do not use a room fogger for spraying a surface. The droplets remain in the air indefinitely.
- Never use any pesticide indoors that requires a gas mask or respirator for its application.
- Never spray entire walls, ceilings, or floors with a pesticide.
- If you're putting out baits, remember that these may be highly toxic. Set them where children and pets can't get at them.

Using Room Foggers Safely

- Cover all food and cooking utensils.
- Close all windows and doors before spraying.
- Leave the room and close the door immediately after releasing the spray.
- Don't reenter the room until the time indicated on the label has elapsed.

Wash Skin and Clothing

Whether eliminating pests chemically indoors or outdoors, always wash any part of your skin that's been exposed. If the chemical has fallen on your clothing, change clothes right away. Don't wear the garments again until they've been laundered. If they've been heavily drenched, wear rubber gloves to pick them up. Put them in a plastic bag and throw them out with the trash.

Storing Pesticides Safely

Buy only enough pesticide for the short term. Store as little of these materials as possible. People may follow all the safety rules when using a pesticide and then stow it under the kitchen sink or with the medicines. Keep chemicals in a locked cabinet and always out of children's view and reach. Check the containers from time to time to make sure they're not leaking.

Application equipment should also be securely locked away or hidden and labeled with the name of the chemical you use it for. Never use such equipment for more than one pesticide.

Pesticide Disposal

According to the University of Massachusetts Cooperative Extension Service, you have three options in disposing of pesticides: Send them to a landfill; bury them diluted on your property; bury them undiluted on your property. *All are undesirable. None of these methods is acceptable for substances labeled "Danger—Poison," or for more than a few pounds or quarts of toxic materials.*

Landfill Alternative

Send only a few containers at a time. Do not puncture any that are pressurized. Wrap each container in several layers of newspaper so that it doesn't rupture in the trash truck and endanger the collectors. For large amounts such as accumulate during a community cleanup, hire a licensed pesticide hauler.

Never burn pesticide containers: You could release poisonous vapors.

If you don't have trash disposal service, crush metal containers, break glass bottles and take them to the public dump; or bury them at least 18 inches deep where they won't contaminate any water source.

Home Burial

Dilute and Bury

This method is suitable for rural areas. Dilute leftover spray at about forty parts of water to one part of pesticide. For example, mix $\frac{1}{2}$ quart of leftover spray with 5 gallons of water. Pour this into a shallow trench in flat, well-drained woods lots or fallow fields with enough soil to absorb the liquid. Avoid sandy areas; bedrock and steep slopes with runoff; and areas near water, gardens, or farms.

Bury Undiluted

Choose a site that will not be disturbed by excavation or construction for at least 5 years. Dig a hole 18 to 20 inches deep. Put in 2 to 3 inches of charcoal followed by 2 to 3 inches of lime. Then pour in the pesticide. Do not bury unopened containers. (See above for safe container disposal.) The chemical must have soil contact to decompose. Cover the pesticide with at least 12 inches of clean soil.

Personal Safety

Spills

If you spill some pesticide while working, stop immediately. Assuming that none of the chemical has splashed you, clean the material up right away. Sweep spilled dusts or powders into a container and dispose of it safely (see previous section). If you've spilled a liquid, soak it up with sand, sawdust, or cat litter. After it's been absorbed, shovel it into a sturdy container and dispose of it as above. When washing spill areas don't allow the wash water to flow into streams, ponds, or other bodies of water. If you're unsure how to dispose of a spilled pesticide, call your county agent or local health department.

Accidental Ingestion

You can be contaminated by a pesticide through your mouth, your respiratory system, your broken or unbroken skin, and your eyes.

Oral Exposure

The most dangerous situation is getting some of the material in your mouth. There it can be absorbed through the tissues without ever reaching the rest of your digestive system. *Don't wait for symptoms to develop.* Get medical help as soon as possible. Take the pesticide container with you so that the chemical can be properly identified for correct treatment.

Eye Exposure

Get to a physician as soon as you can but in the meantime start first aid right away. Hold the eyelid open and wash the eye for at least 15 minutes with a gentle stream of clear, cool running water. Use no chemicals or drugs.

Respiratory Exposure

Go to a well-ventilated area at once and don't go back to the work site until it is clear.

Skin Exposure

Wash thoroughly with soap and water. In both skin and lung exposures, if symptoms develop, get medical help. Again, take the container of pesticide with you.

Poisoning Symptoms

With all the chemical substances we take into our homes (the EPA estimates that the average household buys eighty different ones every year), every adult should be able to recognize common poisoning symptoms. Someone's life may depend on it. Here are the symptoms of pesticide poisoning:

Mild Poisoning: Fatigue, headache, dizziness, blurred vision, heavy sweating, nausea, vomiting, stomachache, diarrhea, possibly heavy salivation.

Moderate Poisoning: Above symptoms worsen. Victim can't walk. Has weakness, chest discomfort, muscle twitching, severe pupil constriction. Salivation increases to point of near drowning.

Severe Poisoning: Unconsciousness, difficulty in breathing, convulsions, in addition to a worsening of the symptoms above. If untreated, death ensues.

Other signs of poisoning, not necessarily from pesticides, are:

An open chemical container

Chemical odor on the breath

Abnormal behavior

Hyperactivity or drowsiness

Shallow or difficult breathing

Burns on the mouth

Loss of consciousness

Poisoning in Children

Poisoning is much harder to spot in toddlers. Anything, says pediatrician Sue Tully, can be a symptom. Be suspicious if your child *suddenly* becomes ill without having had a cold or "the general lousies" for a few days. Or if a youngster *suddenly* starts throwing up. Unlike stomach flu, in poisoning there will be more vomiting than diarrhea. If there are burns around the mouth, or the child begins bumping into things, or is unusually sleepy, *get in touch with a pediatrician immediately.*

What to Do When You Suspect Poisoning

Perhaps what *not* to do is more important than what you should do. "Don't waste time with home remedies like milk or salt water," says Dr. Bernard Portnoy. "The most important thing to do is to get the victim to an emergency room as quickly as possible. Barring that, call a doctor."

Although there are poison information centers around the country, they may not want to talk to individual parents. A doctor is the one to contact them.

The most dangerous pesticides as far as children are concerned are those with the cholinesterase inhibitors, that is, those that affect the central nervous system. A child who has ingested one of these pesticides needs the fastest treatment as well as the most extensive, says Dr. Portnoy.

Don't try to induce vomiting unless a doctor tells you to. If the child has burns around the mouth or is drowsy, vomiting can add to the damage. However, every home with children should have syrup of ipecac on hand, *to be given only if the doctor or poison information center advises it.*

Antidotes Listed on Labels and First-Aid Charts

These should be used only in a pinch, when you can't contact any better source of advice. The problem with the antidotes listed on packages is that they may be completely out of date, especially if you've had the product around a long time. In a random sample taken in 1982, the

New York Poison Control Center found that 85 percent of toxic substances in 1000 households had inadequate or totally wrong antidote information.

The common practice of inducing vomiting with salt water, frequently recommended, is particularly dangerous. Two tablespoons of salt can cause brain and lung damage as well as seizures in a small child. According to Dr. Barry Rumack, director of the Rocky Mountain Poison Center in Denver, thirty to forty children have died from salt poisoning in the past 10 years.

Antipest Gadgets

A number of devices are being promoted in this age of high tech as effective pest eradicators. Don't rush out and buy one and think you have the ultimate answer to all your pest problems. You don't.

Ultrasonic Repellers

Claims for these are very enticing, promising complete pest control without chemicals and without effort. Boasts one, "Rids home of fleas, rodents, roaches, and many other common pests. No effect on humans, cats, dogs, birds or other electronic devices. Proven effective."

The one verifiable fact behind these devices is that a mouse can be killed by ultrasound generated at extremely high frequencies and decibels, both dangerous to human beings. It is very hard to kill a rat this way. In addition, such high frequencies use a great deal of energy.

No Long-Term Effect

As reported in the journal *International Pest Control,* one or two commercial ultrasonic repellers did initially repel rodents slightly. However, the rats being observed soon became accustomed to the noise and ignored it.

High frequencies dissipate rapidly in air so that to produce any repellency at all, a great many units are required. (They sell at anywhere from $15 to $80 each.) In their brief initial effectiveness, the devices tested worked in a space only about ½-meter square and had no effect beyond that area. Yet the manufacturers claimed that they kept rodents out of an area 330 meters square. In addition, ultrasound does not penetrate solid objects, not even the thinnest of barriers, so rats in their harborages are completely unaffected. The following, quoted by the Centers for Disease Control in their manual on rodent control, sums up science's evaluation of ultrasound as pest control:

> "Ultrasound will not drive rodents from building or areas; will not keep them from their usual food supplies and cannot be generated intensely enough to kill rodents in their colonies. Ultrasound has several disadvantages: it is expensive, it is directional and produces "sound shadows" where rodents are not affected and its intensity is rapidly diminished by air and thus of very limited value."

The technology has also been shown to have no effect on cockroaches. Like mice, they'll nest in ultrasound generators.

Electrocuting Devices

A number of electric devices, variously called Fli-Shock, Bugwacker, Fly Zapper, and the like are on the market, claiming they will "destroy flies, mosquitoes, hornets, midges, and other annoying flying insects." These appliances do kill insects. Their jarring crackle on the summer air tells you that from time to time some unlucky flier has been fried. But the gadgets set up a catch-22 situation because they kill as many beneficial insects as pests. Lacewings and lady beetles, those efficient consumers of aphids and other soft-bodied plant suckers, are often electrocuted. The appliances also attract more insects to your garden than you would otherwise have; and they catch more night fliers, mostly harmless moths, than daytime insects.

Worthless against Mosquitoes

If you buy an insect electrocutor hoping to kill mosquitoes with it, you're wasting your money. Scientific tests have shown that humans in the vicinity are consistently more attractive to mosquitoes than the light in the appliance.

Electromagnetic Repellers

Various electromagnetic insect and rodent repellers, some selling for as much as $1000, have been tested and found worthless. When one device was tried that claimed to generate sound waves mosquitoes don't like, nearly 89 percent of the confined mosquitoes took blood whether the machine was turned off or on.

Such devices had no effect on insects' or rodents' eating, drinking, or breeding habits. And in two cases in the field, termites built runways on the gadgets.

Future Pest Controls

Valueless as they are, these technologies may be the shadowy forerunners of more powerful future methods. Control techniques that show some promise are:

- Infrared light traps for specific pests.
- Radio frequency energy that kills stored-food pests without damaging the food.
- Devices that jam pest animals' communications signals.
- Organic pesticides that use the sun's ultraviolet rays to kill insect eggs and larvae. These will not be a panacea for household pests, however, because many of such pests flee light.

All are in the early testing stages, with many technical and economic problems to overcome.

Our Cowboy Mentality Hurts Us

"Americans," says Larry Newman, a vice president of S. C. Johnson, manufacturer of Raid, "have a frontier mentality. The can of spray is their gun. They want the bug to die in front of them." Entomologists, on the other hand, believe that a poison the pest carries back to its cohorts will wipe out a colony much more surely than a product that simply kills bugs on contact. Consumers, however, don't like to see an insect scamper away, so laboratories search for quick-kill formulas.

Scientists working for pesticide manufacturers claim that the bugs are winning our war with them. Their ability to withstand poisons and pass that immunity on to their offspring is a powerful defense, one that stumps researchers over and over. Raid, perhaps the most widely distributed household pesticide in the United States, has had its formula revised twenty-nine times since it was introduced in 1965.

Homes are Dirtier

With the flood of cleaning products and appliances available to U.S. homemakers, one would think that our homes would be cleaner than in the past. Actually they're dirtier. Women by the millions have moved into the paid job market. With both marriage partners working, there's less time for housekeeping. Dust, clutter, forgotten spills, and carelessly handled food all flavor insect life. To avoid the tedium of housecleaning, we turn to chemicals to do the job. For our own protection, let's handle these chemical "brooms" as infrequently as possible—and then only with the greatest caution.

What about Herbal Controls?

Known since ancient times for their insect-repelling odors, herbal controls have been used seriously since the early nineteenth century. Although they won't do much of a job if housekeeping is sloppy, when combined with good home maintenance, some plants do help hold insects off.

Since 1942 the United States Department of Agriculture's Insect Repellent and Attractant Project in Gainesville, Florida, has tested between 300 and 500 herbal derivatives. A few show up well, says Dr. C. E. Schreck, research entomologist with the project. Among the more effective are oil of spearmint, oil of pennyroyal (a common roadside plant in the eastern states), and oil of citronella. As skin repellents, none work as well as the chemical Deet. In addition, oils of pennyroyal and spearmint can cause skin rashes.

Nor can herbals protect a large space. They have little long-distance effectiveness, says Dr. Schreck, and work about as well as a smudge fire.

Some plants, however, are known to repel particular pests. For example, in the southern United States, people plant wax myrtle around the foundations of their home to drive off fleas that cats carry into crawl spaces. A camphor tree will help keep flies away.

Much work on herbals as insect repellents is going on in India. There, where chemicals are too expensive for widespread use, plants like the garlic-smelling neem tree seem to hold promise for good pest control.

What is a commonsense approach to herbals? If you hear of a plant that has helped a friend control a particular pest that is now bugging you, try using the plant along with the methods recommended in this book. Don't expect miraculous or immediate results, but you could achieve acceptable control.

References

Chapter 1

Ballantine, Bill. *Nobody Loves a Cockroach*. Boston: Little, Brown, 1968.

Bateman, Peter L. G. *Household Pests*. Poole, Dorset, England: Blandford Press, 1979.

Bottrell, Dale. *Integrated Pest Management*. Washington, D.C.: Council on Environmental Quality, 1979.

Carson, Rachel. *Silent Spring*. New York: Fawcett Crest, 1962.

Common Pantry Pests and Their Control. Berkeley, CA: Division of Agricultural Sciences, Univ. of California, 1969. Leaflet 2711.

Dehumidifiers. Informal conversation with Hal Thomas of Kamor Air-conditioning, Pasadena, California. Sept. 21, 1983.

Donaldson, Loraine. *Economic Development*. St. Paul, MN: West Publishing, 1983.

Ebeling, Walter. *Urban Entomology*. Berkeley, CA: Division of Agricultural Sciences, Univ. of California, 1978 (rev.).

Hartnack, Hugo. *202 Common Household Pests of North America*. Chicago: Hartnack, 1939.

Horowitz, Joy. "Common Pesticides: A Silent, Deadly Danger." *Los Angeles Times*. Apr. 29–May 1, 1981.

Dangerous Levels of Pesticides Found in Shellfish off Southern California Coast. (Newscast) Los Angeles: KNXT (CBS), Sept. 20, 1982.

Levy, Phil. "Storing Grain: How to Keep the Bugs Out." *The Talking Food Newsletter*. Apr. 1981.

Madon, Minoo. Biologist. California State Department of Health Services. Interview. Oct. 3, 1983.

Dr. Sue Tully. Director of Pediatrics Emergency Room, Los Angeles County–Univ. of Southern California Medical Center. Interview. Sept. 29, 1983.

Williams, Gene B. *The Homeowner's Pest Extermination Handbook*. New York: Arco, 1978.

Chapter 2

Family Medical Guide. The American Medical Association. New York: Random House, 1982.

Hartnack, Hugo. *202 Common Household Pests of North America*. Chicago: Hartnack, 1939.

Horowitz, Joy. "Common Pesticides: A Silent, Deadly Danger." *Los Angeles Times*. Apr. 29–May 1, 1981.

New Columbia Encyclopedia. New York, London: Columbia Univ. Press, 1975.

Olkowski, Helga; Bill Olkowski; Tom Javits; et al. *The Integral Urban House*. San Francisco: Sierra Club Books, 1978.

Scott, H. G. and J. M. Clinton. "An Investigation of 'Cable Mite' Dermatitis." *Annals of Allergy* 25: 409–414, 1967.

Shapiro, David. Director of Orange County [California] Poison Control Center. Interview. June 6, 1981.

Waldron, W. G. "The Role of the Entomologist in Delusory Parasitosis (entomophobia.)" *Enomological Society of America Bulletin* 8: 81–83, 1962.

Wilson, J. W. and H. E. Miller. "Delusion of Parasitosis (acarophobia)." *Archives of Dermatology* 54: 39–56, 1946.

Chapter 3

Ballantine, Bill. *Nobody Loves a Cockroach*. Boston: Little, Brown, 1968.

Boraiko, Allen. "The Indomitable Cockroach." *National Geographic*. Vol. 159, no. 1. Jan. 1981, pp. 130–142.

"Boric Acid for Cockroach Control." Berkeley, CA: Division of Agricultural Sciences, Univ. of California. One-sheet answer #206.

"Cockroach Management." *The IPM Practitioner* 2 (2), Feb. 1980.

Controlling Household Pests. Washington, D.C.: United States Department of Agriculture, Home Garden Bulletin 96, 1971 (rev.).

De Long, D. M. "Beer Cases and Soft Drink Cartons as Insect Distributors." *Pest Control* 30 (7): pp 14–18, 1962.

Ebeling, Walter. *Urban Entomology*. Berkeley, CA: Division of Agricultural Sciences, Univ. of California, 1978 (rev.).

Hartnack, Hugo. *202 Common Household Pests of North America*. Chicago: Hartnack, 1939.

Maas, Rod. Inspector, Pasadena (California) Fire Department. Informal conversation, April 2, 1983.

Meyers, Charles. Biologist, California State Department of Health Services, Interview, May 24, 1983.

Olkowski, Helga; Bill Olkowski; Tom Javits; et al. *The Integral Urban House*. San Francisco: Sierra Club Books, 1979.

Rachesky, Stanley. *Getting Bugs to Bug Off*. New York: Crown, 1978.

Ray, Corinne. Administrator of the Los Angeles County Medical Association Regional Poison Information Center. Interview, June 1, 1981.

Slater, Arthur J. *Controlling Household Cockroaches*. Berkeley, CA: Division of Agricultural Sciences, Univ. of California. Leaflet 21035.

Williams, Gene B. *The Homeowner's Pest Extermination Handbook*. New York: Arco, 1978.

Chapter 4

Ballantine, Bill. *Nobody Loves a Cockroach*. Boston: Little, Brown, 1968.

Bateman, Peter L. G. *Household Pests*. Poole, Dorset, England: Blandford Press, 1979.

Bush-Brown, James and Louise Bush-Brown. *American's Garden Book*. New York: Charles Scribner's Sons, 1947.

Common Pantry Pests and Their Control. Berkeley, CA: Division of Agricultural Sciences, Univ. of California, 1969, Leaflet 2711.

Ebeling, Walter. *Urban Entomology*. Berkeley, CA: Division of Agricultural Sciences, Univ. of California, 1978 (rev.).

Flint, Mary and Robert van den Bosch. *Introduction to Integrated Pest Management*. New York: Plenum, 1981.

Foster, Boyd. President, Arrowhead Mills, Hereford, Texas. Interview. Oct. 31, 1983.

Hartnack, Hugo. *202 Common Household Pests of North America*. Chicago: Hartnack, 1939.

Home and Garden Information. Teletips. Los Angeles, CA: Univ. of California Cooperative Extension Service.

Olkowski, Helga; Bill Olkowski; Tom Javits; et al. *The Integral Urban House*. San Francisco: Sierra Club Books, 1979.

Philbrick, Helen and John Philbrick. *The Bug Book*. Charlotte, VT: Garden Way, 1974.

Pratt, Harry D. *Mites of Public Health Importance and Their Control*. Atlanta: Centers for Disease Control Homestudy Course 3013-G, Manual 8B.

Pratt, Harry D; Kent S. Littig; Harold George Scott. *Household and Stored-Food Insects of Public Health Importance and Their Control*. Atlanta: Centers for Disease Control Homestudy Course 3013-G, Manual 9.

Rachesky, Stanley. *Getting Pests to Bug Off*. New York: Crown, 1978.

Williams, Gene B. *The Homeowner's Pest Extermination Handbook*. New York: Arco, 1978.

Chapter 5

Biological Factors in Domestic Rodent Control. Atlanta: Centers for Disease Control Homestudy Course 3013-G, Manual 10.

Brooks, J. E. "A Review of Commensal Rodents and Their Control." *Critical Reviews in Environmental Control* 3(4): pp. 405–453. 1973.

———. "Roof Rats in Residential Areas—The Ecology of Invasion." *California Vector Views* 13 (9): pp. 69–73. Sept. 1966.

——— and A. M. Bowerman. "Anticoagulant Resistance in Rodents in the United States and Europe." *Journal of Environmental Health*. 37 (6): pp. 537–542.

Canby, Thomas. "The Rat: Lapdog of the Devil." *National Geographic*, vol. 152, no. 1: pp. 60–87. July 1977.

Dutson, Val J. "Use of the Himalayan Blackberry, Rubus Discolor, by the Roof Rat, Rattus Rattus, in California." *California Vector Views*, 20 (8): pp. 59–68. Aug. 1973.

Ebeling, Walter. *Urban Entomology*. Berkeley, CA: Division of Agricultural Sciences, Univ. of California, 1978 (rev.).

Greaves, J. H.; B. D. Rennison; R. Redfern. "Warfarin Resistance in the Ship Rat in Liverpool." *International Pest Control* 15(2): p. 17, 1973.

Hartnack, Hugo. *202 Common Household Pests of North America*. Chicago: Hartnack, 1939.

The House Mouse: Its Biology and Control. Berkeley, CA: Division of Agricultural Sciences, Univ. of California. 1981 (rev.). Leaflet 2945.

"IPM for Rats in Urban Areas." *Urban Ecosystem Management*, from the *IPM Practitioner* 2 (3), Mar. 1980.

Jamieson, Dean. "A History of Roof Rat Problems in Residential Areas of Santa Clara County, California." *California Vector Views* 12 (6): pp. 25–28, June 1965.

Mackie, Richard A. "Control of the Roof Rat, Rattus Rattus, in the Sewers of San Diego." *California Vector Views* 11 (2): pp. 7–10, Feb. 1964.

Philbrick, Helen and John Philbrick. *The Bug Book*. Charlotte, VT: Garden Way, 1974.

Plague: What You Should Know About It. Berkeley, CA: Division of Agricultural Sciences, Univ. of California. Leaflet 21233.

Pratt, Harry D.; Bayard F. Bjornson; Kent S. Littig. *Control of Domestic Rats and Mice.* Atlanta: Centers for Disease Control, Homestudy Course 3013-G, Manual 11.

The Rat: Its Biology and Control. Berkeley, CA: Division of Agricultural Sciences, Univ. of California, 1981 (rev.). Leaflet 2896.

Rats and Mice as Enemies of Mankind. London: British Museum (Natural History). Economic Series 15 (4th ed.).

"Released Laboratory Rats as Community Pests in California." *California Vector Views* 18 (10): pp. 65–68, Oct. 1971.

"Rodent Control Prior to the Closing of Dumps." *California Vector Views* 18 (12): pp. 77–80, Dec. 1971.

"Pets: Toy Fox Terrier." *Los Angeles Times.* Oct. 16, 1983.

Vertebrate Pests: Problems and Control. Ed. by the Subcommittee on Vertebrate Pests, National Research Council, National Academy of Sciences, Washington, D.C., 1970.

Chapter 6

Brown, Paul; Walter Wong; Imre Jelenfy. "A Survey of the Fly Production from Household Refuse Containers in Salinas, California." *California Vector Views* 17 (4): pp. 19–22. Apr. 1970.

Chapman, John and Dean H. Ecke. "Study of a Population of Chironomid Midges Parasitized by Mermithid Nematodes." *California Vector Views* 16 (9): pp. 83–84. Sept. 1969.

Ebeling, Walter. *Urban Entomology.* Berkeley, CA: Division of Agricultural Sciences, University of California, 1978 (rev.).

Ecke, Dean H. and Donald Linsdale. "Control of Green Blow Flies by Improved Methods of Residential Refuse Storage and Collection." *California Vector Views* 14 (4): pp. 19–27. Apr. 1967.

Evans, Howard Ensign. *Life on a Little-Known Planet.* New York: E. P. Dutton, 1968.

Loomis, Edmond C. and Andrew S. Deal. "Control of Domestic Flies." Berkeley, CA: Division of Agricultural Sciences, University of California, May 1975 (rev.). Leaflet 2504.

Madon, Minoo and Charles Meyers. Biologists. Vector Biology and Control Branch, California Department of Health Services. Interview, Oct. 6, 1983.

Magy, H. I. and R. J. Black. "An Evaluation of the Migration of Fly Larvae from Garbage Cans in Pasadena, California." *California Vector Views* 9: pp. 55–59. 1962.

Olkowski, Helga; Bill Olkowski; Tom Javits; et al. *The Integral Urban House.* San Francisco: Sierra Club Books, 1979.

Pratt, Harry D.; Kent S. Littig; Harold George Scott. *Flies of Public Health Importance and Their Control.* Atlanta: Centers for Disease Control Homestudy Course 3013-G, Manual 5.

Ray, Corinne. Director, Los Angeles Medical Association Poison Information Center. Interview. June 1, 1981.

Schreck, C. E. Research entomologist, United States Department of Agriculture, Insect Repellent and Attractant Project, Gainesville, FL. Interview. Apr. 4, 1984.

Spencer, T. S.; R. J. Shimmin; R. F. Schoeppner. "Field Tests of Repellents against the Valley Black Gnat." *California Vector Views* 22 (1): pp. 5–7. Jan. 1975.

Von Frisch, Karl. *Ten Little Housemates*. Trans. Margaret D. Senft. Oxford, England: Pergamon Press. 1960.

Chapter 7

Amonkar, S. V. and A. Banerji. "Isolation and Characterization of the Larvicidal Principal of Garlic." *Science* 174: pp. 1343–1344. 1971.

Ballantine, Bill. *Nobody Loves a Cockroach*. Boston: Little, Brown, 1968.

Biology and Control of Aedes Aegypti. Vector Topics No. 4. Atlanta: Centers for Disease Control, Bureau of Tropical Diseases, Sept. 1979.

Bottrell, Dale. *Integrated Pest Management*. Washington, D.C.: Council on Environmental Quality, 1979.

Brody, Jane E. "Mosquitoes: Why These Pests Find Us So Tasty and How to Stop Them." *Los Angeles Herald Examiner,* July 17, 1983.

Brown, A. W. A. "Attraction of Mosquitoes to Hosts." *Journal of the American Medical Association* Vol. 196: pp. 249–252, 1966.

Carpenter, Stanley and Paul Gieke. "Distribution and Ecology of Mountain Aedes Mosquitoes in California." *California Vector Views* 21, (1–3): pp. 1–8. Jan.–Mar. 1974.

Challet, Gilbert. Director, Orange County Vector Control Agency. Conversation, Oct. 10, 1983.

Control of Western Equine Encephalitis. Vector Topics No. 3. Atlanta: Centers for Disease Control, Bureau of Tropical Diseases, Oct. 1978.

Ebeling, Walter. *Urban Entomology*. Berkeley, CA: Division of Agricultural Sciences, Univ. of California, 1978 (rev.).

Evans, Howard Ensign. *Life on a Little-known Planet*. New York: E. P. Dutton, 1968.

Garcia, Richard; Barbara Des Rochers; and William G. Voigt. "Evaluation of Electronic Mosquito Repellers under Laboratory and Field Conditions," *California Vector Views* 23 (5, 6): pp. 21–23. May–June 1976.

"Genetically Engineered Vaccines Aim at Blocking Infectious Diseases in Millions." *The Wall Street Journal*. Oct. 25, 1983.

Gray, Harold F. and Russel E. Fontaine. "A History of Malaria in California." Proceedings and Papers of the Twenty-fifth Annual Conference of the California Mosquito Control Association. Turlock, California, 1957.

Gutierrez, Michael C.; Ernst P. Zboray; and Patricia A. Gillies. "Insecticide Susceptibility of Mosquitoes in California." *California Vector Views* 23 (7, 8): pp. 27–30. July–August 1976.

Kramer, Marvin C., Executive Director, California Mosquito Control Association. Letter, Oct. 19, 1983.

Madon, Minoo and Charles Meyers. Biologists. Vector Control and Biology Branch, California Department of Health Services. Interview, Oct. 6, 1983.

"Malaria Spreads in New, Deadlier Forms." *Los Angeles Times,* Dec. 25, 1983.

"Mosquitoes' Habits Bug Entomologists." *Los Angeles Times,* Apr. 10, 1983.

Mosquitoes of Public Health Importance and Their Control. Atlanta: Centers for Disease Control Homestudy Course 3013-G, Manual 6.

Mulhern, Thomas. "An Approach to Comprehensive Mosquito Control," *California Vector Views* 19 (9): pp. 61–64. Sept. 1972.

Nasci, Roger S.; Cedric W. Harris; Cyresa K. Porter. "Failure of an Insect Electrocuting Device to Reduce Mosquito Biting." *Mosquito News* 43 (2): pp. 180–184. June 1983.

New Columbia Encyclopedia. New York: Columbia Univ. Press, 1975.

Prevention and Control of Vector Problems Associated with Water Resources. Ft. Collins, CO. Centers for Disease Control, Water Resources Branch, 1965.

Reeves, E. L. and Garcia, C. "Mucilaginous Seeds of the Cruciferae Family as Potential Biological Control Agents for Mosquito Larvae." *Mosquito News* 29: pp 601–607, 1969.

Soares, George; Kevin Hackett; William Olkowski. "IPM for Mosquitoes." *Urban Ecosystem Management* in *The IMP Practitioner* 2 (6). June 1980.

Stewart, James M. Atlanta: Centers for Disease Control, Center for Infectious Diseases. Interview, Oct. 13, 1983.

Sunset Western Garden Book. Menlo Park, CA: Lane Magazine and Book Company, 1973 (rev.).

Van den Bosch, Robert. *The Pesticide Conspiracy.* Garden City, NY: Doubleday, 1978.

Zimmerman, David. "The Mosquitoes Are Coming—and They Are among Man's Most Lethal Foes." *Smithsonian* 14 (3): pp. 28–38. June 1983. Follow-up letter in *Smithsonian* 14 (5): p. 12. Aug. 1983.

Chapter 8

Baker, Norman F. and Thomas B. Farver. "Failure of Brewer's Yeast as a Repellent to Fleas on Dogs." *Journal of the American Veterinary Medicine Association* 2: pp. 212–214. July 15, 1983.

Ballantine, Bill. *Nobody Loves a Cockroach.* Boston: Little, Brown, 1968.

Belfield, Wendell O. and Martin Zucker. *The Very Healthy Cat Book.* New York: McGraw-Hill, 1983.

Bottrell, Dale. *Integrated Pest Management.* Washington, D.C.: Council on Environmental Quality, 1979.

Buck, William B. "Clinical Toxicosis Induced by Insecticides In Dogs and Cats." *Veterinary Medicine/Small Animal Clinician:* p. 1119. Aug. 1979.

"The California Flea Cycle (summer only)." Moraga, CA. Calfornia Veterinary Medical Association. Leaflet.

Clarke, Anna P., DVM. "Pet Question and Answer" *Los Angeles Times.* Aug. 28, 1983.

"Common Ticks Affecting Dogs." Berkeley, CA: Division of Agricultural Sciences of the University of California, 1978 (rev.). Leaflet 2525.

Ebeling, Walter. *Urban Entomology.* Berkeley, CA: Division of Agricultural Sciences, Univ. of California, 1978 (rev.).

"Fleas, Fleas, Fleas." Miami, FL: Adams Veterinary Research Laboratories. Leaflet.

"Fleas Pursue Prey with Dogged Will." *Los Angeles Times.* July 17, 1983.

Frohbieter-Mueller, Jo. "Drown Those Fleas." *Mother Earth News.* July/August 1983.

Halliwell, Richard E. W. "Ineffectiveness of Thiamine (Vitamin B_I) as a Flea Repellent in Dogs." *Journal of the American Animal Hospital Association* 18: pp. 423–426. May/June 1982.

Hartnack, Hugo. *202 Common Household Pests of North America.* Chicago: Hartnack, 1939.

Herb Products Company, North Hollywood, CA. Informal conversation, Dec. 13, 1983.

Insect Pest Management and Control. Vol. 3, Principles of Plant and Animal Pest Control. Washington, DC: National Academy of Sciences, 1969.

Jones, Terry. Chemist. Interview. Dec. 5, 1983.

Keh, Benjamin. "The Brown Dog Tick." *California Vector Views* 11 (5): pp. 27–31. May 1964.

——— and Allan M. Barnes. "Fleas as Household Pests in California." *California Vector Views* 8 (11): pp. 55–58. Nov. 1961.

Lehane, Brendan. *The Compleat Flea.* New York: Viking, 1969.

New Columbia Encyclopedia. New York and London: Columbia Univ. Press, 1975.

Olkowski, Helga: Bill Olkowski; Tom Javits; et al. *The Integral Urban House.* San Francisco: Sierra Club Books, 1979.

Olkowski, William and Linda Laub. "IPM for Fleas." *Urban Ecosystem Management* from *The IPM Practitioner* 2 (9): Sept. 1980.

Peavey, George, DVM, President of Southern California Veterinary Association. Interview. Aug. 29, 1983.

Portnoy, Bernard, MD, Associate Director of Pediatrics Pavilion of Los Angeles County–Univ. of Southern California Medical Center. Interview. Sept. 28, 1983.

Pratt, Harry D. and Harold E. Stark. *Fleas of Public Health Importance and*

Their Control. Atlanta: Centers for Disease Control Homestudy Course 3013-G, Manual 7-A.

Pratt, Harry D.; Kent S. Littig, *Ticks of Public Health Importance and Their Control*. Atlanta: Centers for Disease Control Homestudy Course 3013-G, Manual 8A.

Rachesky, Stanley. *Getting Pests to Bug Off*. New York: Crown, 1978.

Schreck, C. E., Research Entomologist. United States Department of Agriculture Insect Repellent and Attractant Project. Interview, Mar. 4, 1984.

Stewart, James, Entomologist. Centers for Disease Control. Interviews on various occasions in 1983 and 1984.

Von Frisch, Karl. *Ten Little Housemates*. Trans. Margaret D. Senft. Oxford, England: Pergamon Press, 1960.

Wellborn, Stanley. "It's Spring and Insects Are on the March." *U.S. News and World Report*. May 28, 1984.

White, Dee. Conversation. Dec. 6, 1983.

Williams, Gene B. *The Homeowner's Pest Extermination Handbook*. New York: Arco, 1978.

Chapter 9

Ackerman, A. B., "Crabs—The Resurgence of Phthirus Pubis." *New England Journal of Medicine*. 278: 950–51, 1968.

Bottrell, Dale. *Integrated Pest Management*. Washington, D.C.: Council on Environmental Quality, 1979.

"Carpet Beetles, Clothes Moths, Bedbugs, Fleas." East Lansing, MI: Cooperative Extension Service, Michigan State Univ., 1969. Pamphlet.

Ebeling, Walter. *Urban Entomology*. Berkeley, CA: Division of Agricultural Sciences, Univ. of California, 1978 (rev.).

Hartnack, Hugo. *202 Common Household Pests of North America*. Chicago: Hartnack, 1939.

"Head Lice: An Alternative Management Approach." Berkeley, CA: John Muir Institute for Environmental Studies, Center for the Integration of Applied Science. Leaflet.

Howlet, F. M. "Notes on Head and Body Lice and upon Temperature Reactions of Lice and Mosquitoes." *Parasitology* 10: 186–88, 1917.

Insect-Pest Management and Control. Vol. 3 of Principles of Plant and Animal Pest Control. Washington, D.C.: National Academy of Sciences, 1969.

Keh, Benjamin. "Answers to Some Questions Frequently Asked about Pediculosis." *California Vector Views* 26 (5, 6): pp. 51–62, May—June 1979.

———— and John Poorbaugh. "Understanding and Treating Infestations of Lice on Humans." *California Vector Views* 18 (5): pp. 23–30. May 1971.

"Lice and Their Control." Berkeley, CA: Univ. of California Division of Agricultural Sciences. Leaflet 2528.

New Columbia Encyclopedia. New York and London: Columbia Univ. Press, 1975.

Nuttall, G. H. F., "The Biology of Pediculus Humanus." *Parasitology* 10: 18–85, 1917.

Olkowski, Helga; William Olkowski. "Using IPM Principles for Managing Head Lice." *Urban Ecosystem Management in The IPM Practitioner* (1), Jan. 1980.

Olkowski, Helga; William Olkowski; Tom Javits; et al. *The Integral Urban House.* San Francisco: Sierra Club Books, 1979.

Pratt, Harry D.; Kent S. Littig. *Lice of Public Health Importance and Their Control.* Atlanta: Centers for Disease Control, Homestudy Course 3013G, Manual 7-B.

Rachesky, Stanley. *Getting Pests to Bug Off.* New York: Crown, 1978.

"'Tis the Season for Big Trouble from Tiny Lice." *Los Angeles Times.* Oct. 16, 1983.

Wright, W. H. *The Bedbug: Its Habits and Life History and Methods of Control.* Washington, D.C.: Public Health Reports of the United States Public Health Service. Supplement no. 175. 1944.

Chapter 10

"Carpet Beetles." London: British Museum (Natural History). Economic Leaflet no. 8. 1967.

Ebeling, Walter. *Urban Entomology.* Berkeley: Division of Agricultural Sciences, Univ. of California, 1978 (rev.).

Flint, Mary; Robert Van den Bosch. *Introduction to Integrated Pest Management.* New York and London: Plenum, 1981.

Gradidge, J. M. G.; A. D. Aitken; P. E. S. Whalley. "Clothes Moths and Carpet Beetles." London: British Museum (Natural History), Economic Series no. 14. 1967.

Hartnack, Hugo. *202 Common Household Pests of North America.* Chicago: Hartnack, 1939.

"Insect Nuisances in Stores and Homes." London: British Ministry of Agriculture, Fisheries and Food. Leaflet G D 52. May 1978.

Moore, W. S.; C. S. Koehler; C. S. Davis. "Carpet Beetles and Clothes Moths," Berkeley, CA: Division of Agricultural Sciences, Univ. of California, 1979 (rev.). Leaflet 2524.

Slater, Arthur and Georgia Kastanis. "Silverfish and Firebrats and How to Control Them." Berkeley, CA: Division of Agricultural Sciences, Univ. of California, Leaflet 21001, 1977.

Williams, Gene B. *The Homeowner's Pest Extermination Handbook.* New York: Arco, 1978.

Chapter 11

Ballantine, Bill. *Nobody Loves a Cockroach*. Boston: Little, Brown, 1968.

Bastin, Harold. *Insect Communities*. New York: Roy Publishers, 1957.

Bottrell, Dale. *Integrated Pest Management*. Washington, D.C.: Council on Environmental Quality, 1979.

Code of Federal Regulations, Title 40, Part 171.

Ebeling, Walter. "The Extermax System for Control of the Western Drywood Termite." Research paper published by ETEX, Las Vegas, Nevada. No date given.

———. *Urban Entomology*. Berkeley, CA: Division of Agricultural Sciences, Univ. of California, 1978 (rev.).

Flint, Mary and Robert Van den Bosch. *Introduction to Integrated Pest Management*. New York: Plenum, 1981.

Harris, W. Victor. *Termites, Their Recognition and Control*. New York: Wiley, 1961.

Hartnack, Hugo. *202 Common Household Pests of North America*. Chicago: Hartnack, 1939.

Heier, Albert, Public Information Officer, Environmental Protection Agency, Washington, D.C.: Conversation, Jan. 27, 1984.

Horowitz, Joy. "Common Pesticides: A Silent, Deadly Danger." *Los Angeles Times*. April 29–May 1, 1981.

Insect Pest Managment and Control. Vol. 3 of Principles of Plant and Animal Pest Control. Washington, D.C.: National Academy of Sciences, 1969.

"IPM for Termites." *Urban Ecosystem Management*, from *The IPM Practitioner* 2 (12). Dec. 1980.

LaVoie, Robert E., President of Ace Termite and Pest Control Corporation, Los Angeles. Interview, Feb. 2, 1984.

Maeterlinck, Maurice. *The Life of the White Ant*. New York: Dodd, Mead. 1927.

Mahaney, J. H. "Termites on USNS *Sunnyvale*." *Pest Control* 40(7): 17–18, 36, 38. 1972.

Olkowski, Helga; William Olkowski; Tom Javits; et al. *The Integral Urban House*. San Francisco: Sierra Club Books, 1978.

Prestwich, Glenn D. "Dwellers in the Dark: Termites." *National Geographic*. April 1978, vol. 153, no. 4, pp. 532–546.

Rachesky, Stanley. *Getting Pests to Bug Off*. New York: Crown, 1978.

Rambo, George , Director of Technical Operations, National Pest Control Association. Interview, Feb. 6, 1984.

Rust, Michael A.; Donald A. Reierson. "Termites and Other Wood-infesting Insects." Berkeley, CA: Division of Agricultural Sciences, Univ. of California. Leaflet 2532.

Smith, Virgil K.; H. R. Johnston; Raymond H. Beal. *Subterranean Termites: Their Prevention and Control*. USDA Home and Garden Bulletin no. 64. 1972 (rev.), 1979.

Truman, Lee C. and William L. Butts. "Scientific Guide to Pest Control Operations." *Pest Control Magazine,* 1962.

Wilcox, W. Wayne and David L. Wood. "So You've Just Had a Structural Pest Control Inspection." Berkeley, CA: Division of Agricultural Sciences, University of California. Leaflet 2999. July, 1980.

World Book Encyclopedia, Vol. 16. Chicago: Field Enteprises, 1955.

Chapter 12

Carson, Rachel. *Silent Spring.* New York: Fawcett Crest, 1962.

"Casual Home Invading Pests." East Lansing, MI: Cooperative Extension Service, Michigan State Univ., 1969.

Doutt, R. S. "The Praying Mantis." Division of Agricultural Sciences, Univ. of California. Leaflet 21019.

Dreistadt, Steve. "Gypsy Moth in California . . . Is Carbaryl the Answer?" *CBE Environmental Review.* May–June 1983, p. 3.

"Gypsy Moth in Interstate Moves." Los Angeles County Agriculture Commissioner's staff member. Interview. Mar. 21, 1984.

Home Trends. Long and Foster ©, June 1984, distributed by Herbert Hawkin Realty Co.

Huffaker, Carl B., (ed.) *New Technology of Pest Control.* New York: Wiley, 1980.

Insect-Pest Management and Control. Vol. 3 of Principles of Plant and Animal Pest Control. Washington, D.C.: National Academy of Sciences, 1969.

McDonald, Elvin; Jacqueline Héritean; Francesca Morris. *The Color Handbook of Houseplants.* New York: Hawthorn Books, Wentworth, 1975.

"Mealybugs on Houseplants in the Home Landscape." Berkeley, CA: Division of Agricultural Sciences, Univ. of California. Leaflet 21197.

"Mites in the Home Garden and Landscape." Berkeley, CA: Division of Agricultural Sciences, Univ. of California, Leaflet 21048 (B).

Moore, W. S. and C. S. Koehler. "Aphids in the Home Garden and Landscape." Berkeley, CA: Division of Agricultural Sciences, Univ. of California, Leaflet 21032.

Moore, W. S. and C. S. Koehler. "Earwigs and Their Control." Berkeley, CA: Division of Agricultural Sciences, Univ. of California. Leaflet 21010.

Moore, W. S. and C. S. Koehler. "Sowbugs and Pullbugs." Berkeley, CA: Division of Agricultural Sciences, Univ. of California. Leaflet 21015.

Moore, W. S. and K. A. Hesketh. "Snails and Slugs in the Home Garden." Berkeley, CA: Division of Agricultural Sciences, Univ. of California. Leaflet 2530.

Moore, W. S.; J. C. Profita; C. S. Koehler. "Soaps for Home Landscape Insect Control." *California Agriculture* 33 (6). June 1979.

"News from CBE's Minneapolis–St. Paul Office." *CBE Environmental Review.* May–June 1983, p. 3.

"News You Can Use." *U.S. News and World Report*. Oct. 10, 1983.

Philbrick, Helen and John Philbrick. *The Bug Book*. Charlotte, VT: Garden Way, 1980.

Schwartz, P. H. "Control of Insects on Deciduous Fruits and Tree Nuts in the Home Orchard—Without Insecticides," Washington, D.C.: United States Department of Agriculture, Home and Garden Bulletin 211.

Soares, G. G., Jr. "IPM for the Japanese Beetle." Urban Ecosystem Management, *IPM Practitioner* 2 (8). Aug. 1980.

"Your Pest Control Program: One Safe Step at a Time." *Organic Garden and Farming,* April 1975, vol. 22 (4): pp. 87–91.

West, Ron. "The Backyard Jungle; Spider Mites." *Mother Earth News*. July–August 1983.

"Whiteflies on Outdoor and Indoor Plants." Berkeley, CA: Division of Agricultural Sciences, Univ. of California. Leaflet 21267.

Yepsen, Roger B. Jr., (ed.). *Organic Plant Protection*. Emmaus, PA: Rodale, 1976.

Chapter 13

"Ants Indoors." London: British Ministry of Agriculture, Fisheries, and Food. Leaflet 366.

"Ants and Their Control." Berkeley, CA: Division of Agricultural Sciences, Univ. of California. Leaflet 2526.

Ballantine, Bill. *Nobody Loves a Cockroach*. Boston: Little, Brown, 1967.

Carson, Rachel. *Silent Spring*. New York: Fawcett Crest, 1962.

"Control of Yellowjackets and Similar Wasps." Berkeley, CA: Division of Agricultural Sciences, Univ. of California. Leaflet 2527.

Ebeling, Walter. *Urban Entomology*. Berkeley, CA: Division of Agricultural Sciences, Univ. of California, 1978 (rev.).

Evans, Howard Ensign. *Wasp Farm*. Garden City, NY: Natural History Press, 1963.

——— and Mary Jane West Eberhard. *The Wasps*. Ann Arbor: Univ. of Michigan Press, 1970.

Gertsch, Willis. *American Spiders*. New York: Van Nostrand Reinhold, 1979.

Gorham, J. Richard. "The Brown Recluse." Washington, D.C.: United States Public Health Service. Publication No. 2062.

———. "The Geographic Distribution of the Brown Recluse Spider, Loxosceles Reclusa, and Related Species in the United States." *Cooperative Economic Insect Report (CEIR)* 18: 171–75, 1968.

Hutchins, Ross E. *Insects*. Englewood Cliffs, NJ: Prentice-Hall, 1966.

Keh, Benjamin. "The Black Widow Spider." *California Vector Views* 3 (1): pp. 1, 3–4. Jan. 1956.

——. "A Brief Review of Necrotic Arachnidism or North American Loxoscelism." *California Vector Views* 14 (7): pp. 48–50. July 1967.

——. "Loxosceles Spiders in California." *California Vector Views* 17 (5): pp. 29–34, May 1970.

New Columbia Encyclopedia. New York, London: Columbia Univer. Press, 1975.

Olkowski, Helga; William Olkowski; Tom Javits; et al. *The Integral Urban House.* San Francisco: Sierra Club Books, 1978.

Olkowski, William. "Ants in the Bay Area." John Muir Institute. Memo, 1973.

Philbrick, Helen and John Philbrick. *The Bug Book.* Charlotte, VT: Garden Way, 1974.

Rachesky, Stanley. *Getting Pests to Bug Off.* New York: Crown, 1978.

Stahnke, H. L. *The Treatment of Venomous Bites and Stings.* Tempe: Arizona State University, 1966.

Teale, Edwin Way. *The Strange Lives of Familiar Insects.* New York: Dodd, Mead, 1962.

Von Frisch, Karl. *Ten Little Housemates.* Trans. Margaret D. Senft. Oxford, England: Pergamon Press, 1960.

Waldron, William G.; Minoo B. Madon; Terry Sudderth. "Observations on the Occurrence and Ecology of *Loxosceles Laeta* in Los Angeles County, California." *California Vector Views* 22 (4): pp. 29–35, Apr. 1975.

"Wasps." London: British Ministry of Agriculture, Fisheries, and Food. GD 53.

Williams, Gene B. *The Homeowners' Pest Extermination Handbook.* New York: Arco, 1978.

World Book Encyclopedia. Chicago: Field Enterprises, 1955.

The Yellowjackets of America North of Mexico. Washington, D.C.: United States Department of Agriculture. Agriculture Handbook no. 552, June 1981.

Chapter 14

Aikman, Lonnelle. "Herbs for All Seasons." *National Geographic,* March 1983, vol. 163 (3): pp. 386–393.

Ballantine, Bill. *Nobody Loves a Cockroach.* Boston: Little, Brown, 1968.

Ballard, J. B.; R. F. Gold. "Ultrasonics—No Effect on Cockroach Behavior," *Pest Control,* June 1982.

Bennett, G. W.; E. S. Runstrom; J. A. Wieland. "Pesticide Use in Homes." *Bulletin of the Entomological Society of America.* Spring 1983.

Biological Factors in Domestic Rodent Control. Atlanta: Centers for Disease Control. Homestudy Course 3013-G, Manual 10.

Blondell, Jerome. "Pesticide-Related Injuries Treated in United States Hospital Emergency Rooms." *1981 Calendar Year Report.* Washington, D.C.: Health Effects Branch, Environmental Protection Agency.

Bottrell, Dale. *Integrated Pest Management*. Washington, D.C: Council on Environmental Quality, 1979.

"Children Act Fast—So Do Poisons." Washington, D.C.: United States Consumer Product Safety Commission. Leaflet.

Ebeling, Walter. *Urban Entomology*. Berkeley, CA: Division of Agricultural Sciences, Univ. of California, 1978 (rev.).

"EPA Stops Sale of Several Electromagnetic Insect and Rodent Repellers." Washington, D.C.: News Release of the Environmental Protection Agency (EPA), 1980.

"Family's Dream Dashed by Bitter Irony." *Los Angeles Times*. Mar. 4, 1984.

Greaves, J. H. and F. P. Rowe. "Responses of Confined Rodent Populations to an Ultrasound Generator." *Journal of Wildlife Management* (2): pp. 409–417. April 1969.

Highlights of the Findings of the National Household Pesticide Usage Study, 1976–1977. Washington, DC: EPA Memo.

Howell, H. N., Jr. and T. A. Granovsky. "Report on Evaluation of Ultrasonic Pest Control Devices vs. American Cockroaches." Unpublished. College Station, TX: Department of Entomology, Texas State University.

Insect-Pest Management and Control. Vol. 3 of Principles of Plant and Animal Pest Control. Washington, DC: National Academy of Sciences, 1969.

Kostich, B. E. "Nature's Insecticide." *Life and Health Quarterly,* fourth quarter, 1980.

Kutz, Frederick W. "Evaluation of an Electronic Mosquito Repelling Device." *Mosquito News,* December 1974.

LaVoie, G. K. and J. F. Glahn. "Ultrasound as a Deterrent to Rattus Norvegicus." Denver, CO: Denver Wildlife Research Center, Mar. 1976.

McCarron, Margaret M. "Acute Yellow Phosphorus Poisoning from Pesticide Pastes." *Clinical Toxicology,* vol. 18 (6): pp. 693–711, 1981.

Meehan, A. P. "Attempts to Influence the Feeding Behavior of Brown Rats Using Ultrasonic Noise Generators." *International Pest Control,* pp. 12–15. July–August 1976.

Nasci, R. S.; C. W. Harris; C. K. Porter. "Failure of an Insect Electrocuting Device to Reduce Mosquito Biting." *Mosquito News* 43 (2): pp. 180–184. June 1983.

"Pesticide Disposal for the Homeowner." *Pesticide Facts*. Boston: Cooperative Extension, University of Massachusetts.

Pesticide Toxicities. Berkeley, CA: Division of Agricultural Sciencs, Univ. of California. Leaflet 21062, 1979.

"Poison Antidotes Can Be Lethal Too." *Los Angeles Times*. May 2, 1983.

"Poison Prevention Packaging." Washington, DC: United States Consumer Product Safety Commission. Product Safety Fact Sheet no. 46.

Riker, Tom. *The Healthy Garden Book*. New York: Stein and Day, 1979.

Safe Use of Agricultural and Household Pesticides. Washington, DC: United States Department of Agriculture Handbook no. 321, 1967.

Schreck, C. E. Research entomologist, USDA Insect Repellent and Attractant Project. Interview, Mar. 4, 1984.

Stead, Frank M. "Pesticides in Relation to Environmental Health." *California Vector Views* 11 (1): pp. 1–3, Jan. 1964.

Stimman, M. W.; J. Litewka. "Using Pesticides Safely in the Home and Yard." Berkeley: Division of Agricultural Sciences, Univ. of California. Leaflet 21095, 1979.

"Sunlight that Kills." *Forbes,* May 23, 1983.

Zaslow, Jeffrey. "Scientists Seek Upper Hand in Insect Wars." *The Wall Street Journal,* Mar. 6, 1984.

Index

267

$c. 1$

648.7 Lifton, Bernice.
Lif
        Bugbusters

DATE			

MADISON COUNTY
CANTON PUBLIC LIBRARY SYSTEM
CANTON, MISS. 39046

© THE BAKER & TAYLOR CO.

648.7 Lifton, Bernice.
Lif
       Bugbusters

c. 1

DATE			

MADISON COUNTY
CANTON PUBLIC LIBRARY SYSTEM
CANTON, MISS. 39046

© THE BAKER & TAYLOR CO.